CAMBRIDGE LIBRARY COLLECTION

Books of enduring scholarly value

Technology

The focus of this series is engineering, broadly construed. It covers technological innovation from a range of periods and cultures, but centres on the technological achievements of the industrial era in the West, particularly in the nineteenth century, as understood by their contemporaries. Infrastructure is one major focus, covering the building of railways and canals, bridges and tunnels, land drainage, the laying of submarine cables, and the construction of docks and lighthouses. Other key topics include developments in industrial and manufacturing fields such as mining technology, the production of iron and steel, the use of steam power, and chemical processes such as photography and textile dyes.

Electromagnetic Waves

Oliver Heaviside (1850–1925), the self-taught physicist and electrical engineer, began his career as an operator on the newly laid Anglo-Danish telegraph cable in 1868. The most advanced electrical technology of the time, the cable system inspired several of his early mathematical papers. This monograph, first published as a paper in the *Philosophical Magazine* in 1888, then as a book in 1889, draws on his established work on telegraphic propagation and self-inductance, and on Maxwell's field theory. In a fascinating insight into the contemporary scientific community, he complains that these subjects are still often misunderstood, and explains his formulae afresh from several angles. Also covered – and frequently questioned – are contemporary theories of permittivity, the speed of electromagnetic waves, and the dielectric properties of conductors. Heaviside's *Electrical Papers* (2 volumes, 1892) and his *Electromagnetic Theory* (3 volumes, 1893–1912) have also been reissued in this series.

Cambridge University Press has long been a pioneer in the reissuing of out-of-print titles from its own backlist, producing digital reprints of books that are still sought after by scholars and students but could not be reprinted economically using traditional technology. The Cambridge Library Collection extends this activity to a wider range of books which are still of importance to researchers and professionals, either for the source material they contain, or as landmarks in the history of their academic discipline.

Drawing from the world-renowned collections in the Cambridge University Library, and guided by the advice of experts in each subject area, Cambridge University Press is using state-of-the-art scanning machines in its own Printing House to capture the content of each book selected for inclusion. The files are processed to give a consistently clear, crisp image, and the books finished to the high quality standard for which the Press is recognised around the world. The latest print-on-demand technology ensures that the books will remain available indefinitely, and that orders for single or multiple copies can quickly be supplied.

The Cambridge Library Collection brings back to life books of enduring scholarly value (including out-of-copyright works originally issued by other publishers) across a wide range of disciplines in the humanities and social sciences and in science and technology.

Electromagnetic Waves

Oliver Heaviside

CAMBRIDGE UNIVERSITY PRESS

Cambridge, New York, Melbourne, Madrid, Cape Town,
Singapore, São Paolo, Delhi, Tokyo, Mexico City

Published in the United States of America by Cambridge University Press, New York

www.cambridge.org
Information on this title: www.cambridge.org/9781108041591

© in this compilation Cambridge University Press 2012

This edition first published 1889
This digitally printed version 2012

ISBN 978-1-108-04159-1 Paperback

This book reproduces the text of the original edition. The content and language reflect the beliefs, practices and terminology of their time, and have not been updated.

Cambridge University Press wishes to make clear that the book, unless originally published by Cambridge, is not being republished by, in association or collaboration with, or with the endorsement or approval of, the original publisher or its successors in title.

From the PHILOSOPHICAL MAGAZINE, Vols. 25–27 (1888–89)

ELECTROMAGNETIC WAVES.

BY

OLIVER HEAVISIDE.

LONDON:

PRINTED BY TAYLOR AND FRANCIS, RED LION COURT, FLEET STREET.

1889.

From the PHILOSOPHICAL MAGAZINE for February 1888.

On Electromagnetic Waves, especially in relation to the Vorticity of the Impressed Forces; and the Forced Vibrations of Electromagnetic Systems. By OLIVER HEAVISIDE.

1. *SUMMARY of Electromagnetic Connexions.*—To avoid indistinctness, I start with a short summary of Maxwell's scheme, so far as its essentials are concerned, in the form given by me in January 1885 †.

Two forces, electric and magnetic, **E** and **H**, connected with the three fluxes,—electric displacement **D**, conduction-current **C**, and magnetic induction **B**; thus

$$\mathbf{B}=\mu\mathbf{H}, \quad \mathbf{C}=k\mathbf{E}, \quad \mathbf{D}=(c/4\pi)\mathbf{E}. \quad . \quad . \quad . \quad (1)$$

Two currents, electric and magnetic, **Γ** and **G**, each of which is proportional to the curl or vorticity of the *other* force, not counting impressed; thus,

$$\text{curl}\,(\mathbf{H}-\mathbf{h})=4\pi\mathbf{\Gamma}, \quad . \quad . \quad . \quad . \quad . \quad . \quad (2)$$

$$\text{curl}\,(\mathbf{e}-\mathbf{E})=4\pi\mathbf{G}; \quad . \quad . \quad . \quad . \quad . \quad (3)$$

where **e** and **h** are the impressed parts of **E** and **H**. These currents are also directly connected with the corresponding forces through

$$\mathbf{\Gamma}=\mathbf{C}+\dot{\mathbf{D}}, \quad \mathbf{G}=\dot{\mathbf{B}}/4\pi. \quad . \quad . \quad . \quad . \quad . \quad (4)$$

* See the opening sections of "Electromagnetic Induction and its Propagation," Electrician, Jan. 3, 1885, and after.

An auxiliary equation to exclude unipolar magnets, viz.

$$\text{div.}\,\mathbf{B}=0, \quad . \quad . \quad . \quad . \quad . \quad . \quad . \quad . \quad (5)$$

expressing that **B** has no divergence. The most important feature of this scheme is the equation (3), as a fundamental equation, the natural companion to (2).

The derived energy relations are not necessary, but are infinitely too useful to be ignored. The electric energy U, the magnetic energy T, and the dissipativity Q, all per unit volume, are given by

$$\mathrm{U}=\tfrac{1}{2}\mathbf{ED}, \quad \mathrm{T}=\tfrac{1}{2}\mathbf{HB}/4\pi, \quad \mathrm{Q}=\mathbf{EC}. \quad . \quad . \quad . \quad (6)$$

The transfer of energy **W** per unit area is expressed by a vector product,

$$\mathbf{W}=\mathrm{V}(\mathbf{E}-\mathbf{e})(\mathbf{H}-\mathbf{h})/4\pi, \quad . \quad . \quad . \quad . \quad (7)$$

and the equation of activity per unit volume is

$$\mathbf{eF}+\mathbf{hG}=\mathrm{Q}+\dot{\mathrm{U}}+\dot{\mathrm{T}}+\text{div.}\,\mathbf{W}, \quad . \quad . \quad . \quad . \quad (8)$$

from which **W** disappears by integration over *all* space.

The equations of propagation are obtained by eliminating either **E** or **H** between (2) and (3), and of course take different forms according to the geometrical coordinates selected.

In a recent paper I gave some examples * illustrating the extreme importance of the lines of vorticity of the impressed forces, as the sources of electromagnetic disturbances. Those examples were mostly selected from the extended developments which follow. Although, being special investigations, involving special coordinates, vector methods will not be used, it will still be convenient occasionally to use the black letters when referring to the actual forces or fluxes, and to refer to the above equations. The German or Gothic letters employed by Maxwell I could never tolerate, from inability to distinguish one from another in certain cases without looking very hard. As regards the notation **EC** for the scalar product of **E** and **C** (instead of the quaternion −S**EC**) it is the obvious practical extension of EC, the product of the tensors, what **EC** reduces to when **E** and **C** are parallel.†

* Phil. Mag. Dec. 1887, "On Resistance and Conductance Operators," § 8, p. 487.

† In the early part of my paper "On the Electromagnetic Wave-Surface," Phil. Mag. June 1885, I have given a short introduction to the Algebra of vectors (not quaternions) in a practical manner, *i. e.* without metaphysics. The result is a thoroughly practical working system. The matter is not an insignificant one, because the extensive use of vectors in mathematical physics is bound to come (the sooner the better), and my method furnishes a way of bringing them in without any study of Quaternions (which are scarcely wanted in Electromagnetism, though

2. *Plane Sheets of Impressed Force in a Nonconducting Dielectric.*—We need only refer to impressed electric force **e**, as solutions relating to **h** are quite similar. Let an infinitely extended nonconducting dielectric be divided into two regions by an infinitely extended plane (x, y), on one side of which, say the left, or that of $-z$, is a field of **e** of uniform intensity e, but varying with the time. If it be perpendicular to the boundary, it produces no flux. Only the tangential component can be operative. Hence we may suppose that **e** is parallel to the plane, and choose it parallel to x. Then **E**, the force of the flux, is parallel to x, of intensity E say, and the magnetic force, of intensity H, is parallel to y. Let $e=f(t)$; the complete solutions due to the impressed force are then

$$\mathrm{E}=\mu v\mathrm{H}=-\tfrac{1}{2}f(t-z/v) \quad . \quad . \quad . \quad . \quad . \quad (9)$$

on the right side of the plane, where z is $+$, and

$$-\mathrm{E}=\mu v\mathrm{H}=-\tfrac{1}{2}f(t+z/v) \quad . \quad . \quad . \quad . \quad . \quad (10)$$

on the left side of the plane, where z is $-$. In the latter case we must deduct the impressed force from E to obtain the force of the field, say F, which is therefore

$$\mathrm{F}=-f(t)+\tfrac{1}{2}f\left(t+\frac{z}{v}\right). \quad . \quad . \quad . \quad . \quad (11)$$

The results are most easily followed thus. At the plane itself, where the vortex-lines of **e** are situated, we, by varying e, produce simultaneous changes in H, thus,

$$\mathrm{H}=\frac{e}{2\mu v}, \quad . \quad . \quad . \quad . \quad . \quad . \quad (12)$$

at the plane. This disturbance is then propagated both ways undistorted at the speed $v=(\mu c)^{-\frac{1}{2}}$.

On the other hand, the corresponding electric displacements are oppositely directed on the two sides of the plane.

Since the line-integral of H is electric current, and the line-integral of e is electromotive force, the ratio of e to H is the resistance-operator of an infinitely long tube of unit area; a constant, measurable in ohms, being 60 ohms in vacuum, or 30 ohms on each side. Why it is a constant is simply

---

they may be added on), and allows us to work without change of notation, especially when the vectors are in special type, as they should be, being entities of widely different nature from scalars. I denote a vector by (say) **E**, its tensor by E, and its x, y, z components, when wanted, by E_1, E_2, E_3. The perpetually occurring scalar product of two vectors requires no prefix. The prefix V of a vector product should be a special symbol.

because the waves cannot return, as there is no reflecting barrier in the infinite dielectric.

3. If the impressed force be confined to the region between two parallel planes distant $2a$ from one another, there are now two sources of disturbances, which are of opposite natures, because the vorticity of e is oppositely directed on the two planes, so that the left plane sends out both ways disturbances which are the negatives of those simultaneously emitted by the right plane. Thus, if the origin of z be midway between the planes, we shall have

$$\mathrm{E}=\mu v\mathrm{H}=-\tfrac{1}{2}f\left(t-\frac{z-a}{v}\right)+\tfrac{1}{2}f\left(t-\frac{z+a}{v}\right) \quad . \quad (13)$$

on the right side of the stratum of e, and

$$-\mathrm{E}=\mu v\mathrm{H}=-\tfrac{1}{2}f\left(t+\frac{z+a}{v}\right)+\tfrac{1}{2}f\left(t+\frac{z-a}{v}\right) \quad . \quad (14)$$

on the left side. If therefore e vary periodically in such a way that

$$f(t)=+f(t+2a/v), \quad . \quad . \quad . \quad . \quad . \quad (15)$$

there is no disturbance outside the stratum, after the initial waves have gone off, the disturbance being then confined to the stratum of impressed force.

Decreasing the thickness of the stratum indefinitely leads to the result that the effect due to $e=f(t)$ in a layer of thickness dz at $z=0$ is, on the right side,

$$\mathrm{H}=-\frac{1}{2\mu v}\left\{f\left(t-\frac{z}{v}\right)-f\left(t-\frac{z+dz}{v}\right)\right\}$$

$$=-\frac{c\,dz}{2}f'\left(t-\frac{z}{v}\right), \quad . \quad . \quad . \quad . \quad . \quad . \quad . \quad (16)$$

since $\mu cv^2=1$; on the left side the + sign is required.

We can now, by integration, express the effect due to $e=f(z, t)$, viz.

$$\mathrm{H}=-\frac{c}{2}\int_{-\infty}^{z}\frac{d}{dt}f\left(t-\frac{z-z'}{v}, z'\right)dz'+\frac{c}{2}\int_{z}^{\infty}\frac{d}{dt}\left(t+\frac{z-z'}{v}, z'\right)dz', \quad (17)$$

$$\mathrm{E}=e-\frac{1}{2v}\int_{-\infty}^{z}\frac{d}{dt}f\left(t-\frac{z-z'}{v}, z'\right)dz'-\frac{1}{2v}\int_{z}^{\infty}\frac{d}{dt}\left(t+\frac{z-z'}{v}, z'\right)dz'. \quad (18)$$

In these, however, a certain assumption is involved, viz. that e vanishes at ∞ both ways, because we base the formulæ upon (16), which concerns a layer of e on both sides of which e is zero. Now the disturbances really depend upon de/dz, for there can be none if this be zero. By (12) the elementary

de/dz through distance dz instantly produces

$$\mathrm{H}=\frac{1}{2\mu v}\frac{de}{dz}dz \quad . \quad . \quad . \quad . \quad . \quad . \quad (19)$$

at the place. If, therefore, $e=f(z,t)$, the H solution at any point consists of the positive waves coming from planes of de/dz on the left, producing, say, H_1, and of H_2 due to the negative waves from the planes of de/dz on the right side, making the complete solution

$$\left.\begin{aligned}\mathrm{H}&=\mathrm{H}_1+\mathrm{H}_2,\\ \mathrm{E}&=\mu v(\mathrm{H}_1-\mathrm{H}_2)\;;\end{aligned}\right\} \quad . \quad . \quad . \quad . \quad . \quad . \quad (20)$$

where

$$\mathrm{H}_1=\frac{1}{2\mu v}\int_{-\infty}^{z}\frac{d}{dz'}f\left(t-\frac{z-z'}{v},z'\right)dz', \quad . \quad . \quad (21)$$

$$\mathrm{H}_2=\frac{1}{2\mu v}\int_{z}^{\infty}\frac{d}{dz'}f\left(t+\frac{z+z'}{v},z'\right)dz' \quad . \quad . \quad . \quad (22)$$

This is the most rational form of solution, and includes the case of $e=f(t)$ only. The former may be derived from it by effecting the integrations in (21) and (22); remembering in doing so that the differential coefficient under the sign of integration is not the complete one with respect to z', as it occurs twice, but only to the second z', and further assuming that $e=0$ at infinity.

4. *Waves in a Conducting Dielectric. How to remove the Distortion due to the Conductivity.*—Let us introduce a new physical property into the conducting medium, namely that it cannot support magnetic force without dissipation of energy at a rate proportional to the square of the force, a property which is the magnetic analogue of electric conductivity. We make the equations (2) and (3) become, if $p=d/dt$,

$$\mathrm{curl}\,\mathbf{H}=(4\pi k+cp)\mathbf{E}, \quad . \quad . \quad . \quad . \quad . \quad (23)$$

$$-\mathrm{curl}\,\mathbf{E}=(4\pi g+\mu p)\mathbf{H}\;; \quad . \quad . \quad . \quad . \quad . \quad (24)$$

if there be no impressed force at the spot, where g is the new coefficient of magnetic conductivity, analogous to k.

Let

$$\left.\begin{aligned}&4\pi k/2c=q_1, \quad 4\pi g/2\mu=q_2,\\ &q_1+q_2=q, \quad q_1-q_2=s,\\ &\mathbf{E}=\epsilon^{-qt}\mathbf{E}_1, \quad \mathbf{H}=\epsilon^{-qt}\mathbf{H}_1.\end{aligned}\right\} \quad . \quad . \quad . \quad (25)$$

Substitution in (23), (24) lead to

$$\mathrm{curl}\,\mathbf{H}_1=c(s+p)\mathbf{E}_1, \quad . \quad . \quad . \quad . \quad . \quad . \quad (26)$$

$$-\mathrm{curl}\,\mathbf{E}_1=\mu(-s+p)\mathbf{H}_1. \quad . \quad . \quad . \quad . \quad . \quad (27)$$

If $s=0$, these are the equations of electric and magnetic force in a nonconducting dielectric. If therefore the new g be of such magnitude as to make $s=0$, we cause disturbances to be propagated in the conducting dielectric in identically the same manner as if it were nonconducting, but with a uniform attenuation at a rate indicated by the time-factor ϵ^{-qt}.

5. *Undistorted Plane Waves in a Conducting Dielectric.*—Taking z perpendicular to the plane of the waves, we now have, as special forms of (23), (24),

$$-\frac{d\text{H}}{dz}=(4\pi k+cp)\text{E}, \quad \ldots \ldots \quad (28)$$

$$-\frac{d\text{E}}{dz}=(4\pi g+\mu p)\text{H}, \quad \ldots \ldots \quad (29)$$

E being the tensor of **E**, parallel to x, and H the tensor of **H**, parallel to y, and both being functions of z and t.

Given $\text{E}=\text{E}_0$ and $\text{H}=\text{H}_0$ at time $t=0$, functions of z only, decompose them thus,

$$2f_1=\text{E}_0+\mu v\text{H}_0, \quad \ldots \ldots \quad (30)$$

$$2f_2=\text{E}_0-\mu v\text{H}_0. \quad \ldots \ldots \quad (31)$$

Here f_1 makes the positive and f_2 the negative wave, and at time t the solutions are, due to the initial state, when $s=0$,

$$\text{E}=\epsilon^{-qt}\{f_1(z-vt)+f_2(z+vt)\}, \quad \ldots \quad (32)$$

$$\mu v\text{H}=\epsilon^{-qt}\{f_1(z-vt)-f_2(z+vt)\}. \quad \ldots \quad (33)$$

The only difference from plane waves in a nonconducting dielectric is in the uniform attenuation that goes on, due to the dissipation of energy, which is so balanced on the electric and magnetic sides as to annihilate the distortion the waves would undergo were s finite, whether positive or negative.

6. *Practical Application. Imitation of this Effect.*—When I introduced* the new property of matter symbolized by the coefficient g, it was merely to complete the analogy between the electric and magnetic sides of electromagnetism. The property is non-existent, so far as I know. But I have more recently found how to precisely imitate its effect in another electromagnetic problem, also relating to plane waves, making use of electric conductivity to effect the functions of both k and g in §§ 4 and 5. In the case of § 5, first remove both conductivities, so that we have plane waves unattenuated and undistorted. Next put a pair of parallel wires of no resistance in the dielectric, parallel to z, and let the lines of electric force

* See second footnote, p. 130.

terminate upon them, whilst those of magnetic force go round the wires. We shall still have these plane electromagnetic waves with curved lines of force propagated undistorted and unattenuated, at the same speed v. If V be the line-integral of **E** across the dielectric from one wire to the other, and 4πC be the line-integral of **H** round either wire, we shall have

$$-\frac{dV}{dz} = LpC, \quad . \quad . \quad . \quad . \quad . \quad . \quad . \quad (34)$$

$$-\frac{dC}{dz} = SpV, \quad . \quad . \quad . \quad . \quad . \quad . \quad . \quad (35)$$

(34) taking the place of (29), and (35) of (28), with k and g both zero. Here L and S are the inductance and permittance of unit length of the circuit of the parallel wires, and $v = (LS)^{-\frac{1}{2}}$.

Next let the wires have constant resistance R per unit length to current in them, and let the medium between them be conducting (to a very low degree), making K the conductance per unit length across from one wire to the other. We then turn the last equations into

$$-\frac{dV}{dz} = (R + Lp)C, \quad . \quad . \quad . \quad . \quad (36)$$

$$-\frac{dC}{dz} = (K + Sp)V, \quad . \quad . \quad . \quad . \quad (37)$$

and have a complete imitation of the previous unreal problem. The two dissipations of energy are now due to R in the wires, and to K in the dielectric, it being that in the wires which takes the place of the unreal magnetic dissipation. The relation $R/L = K/S$, which does not require excessive leakage when the wires are of copper of low resistance, removes the distortion otherwise suffered by the waves. I have, however, found that when the alternations of current are very rapid, as in telephony, there is very little distortion produced by copper wires, even without the leakage required to wholly remove it, owing to R/Ln becoming small, $n/2\pi$ being the frequency; an effect which is greatly assisted by increasing the inductance (see Note A, p. 151). Of course there is little resemblance between this problem and that of the long and slowly-worked submarine cable, whether looked at from the physical side or merely from the numerical point of view, the results being then of different orders of magnitude. A remarkable misconception on this point seems to be somewhat generally held. It seems to be imagined that self-induction is harmful* to long-distance telephony. The precise contrary is the case. It is the very life and soul of it, as is proved both by practical experience in America and the Continent on very long copper circuits, and by examining the theory of the matter. I have proved this in considerable detail†; but they will not believe it. So far does the misconception extend that it has perhaps contributed to leading Mr. W. H. Preece to conclude that the coefficient of self-induction in copper circuits is negligible (several hundred times smaller than it can possibly be), on the basis of his recent remarkable experimental researches.

The following formula, derived from my general formulæ‡, will show the *rôle* played by self-induction. Let R and L be the resistance and inductance per unit length of a perfectly insulated circuit of length l, short-circuited at both ends. Let a rapidly sinusoidal impressed force act at one end of amplitude e_0, and let C_0 be the amplitude of the current at the distant end. Then, if the circuit be very long,

$$C_0 = \frac{2e_0}{Lv}\epsilon^{-Rl/2Lv}, \quad . \quad . \quad . \quad . \quad . \quad (38)$$

where v is the speed $(LS)^{-\frac{1}{2}} = (\mu c)^{-\frac{1}{2}}$, provided R/Ln be small, say $\frac{1}{4}$. It may be considerably greater, and yet allow (38) to be nearly true. We can include nearly the whole range of telephonic frequencies by using suspended copper wires of low resistance§.

It is resistance that is so harmful, not self-induction; as, in combination with the electrostatic permittance, it causes immense distortion of waves, unless counteracted by increasing the inductance, which is not often practicable (see Note B, p. 152).

7. *Distorted Plane Waves in a Conducting Dielectric.*—Owing to the fact that, as above shown, we can fully utilize solutions involving the unreal g, by changing the meaning of the symbols, whilst still keeping to plane electromagnetic waves, we may preserve g in our equations (28) and (29), remembering that H has to become C, E become V, $4\pi k$

* W. H. Preece, F.R.S., "On the Coefficient of Self-Induction of Copper Wires," B. A. Meeting, 1887.

† "El. Mag. Ind. and its Propagation," Electrician, Arts. xl. to l. (1887).

‡ See the sinusoidal solutions in Part II. and Part V. of "On the Self-Induction of Wires," Phil. Mag. Sept. 1886 and Jan. 1887.

§ The explanation of the $\frac{1}{2}Lv$ dividing e_0 in (38), instead of the Lv we might expect from the μv resistance-operator of a tube of unit section infinitely long one way only, is that, on arrival at the distant end of the line, the current is immediately doubled in amplitude by the reflected wave. The second and following reflected waves are negligible, on account of the length of the line.

become K, c become S, $4\pi g$ become R, and μ become L, when making the application to the possible problem; whilst, when dealing with a real conducting dielectric, g has to be zero.

Required the solutions of (28) and (29) due to any initial states E_0 and H_0, when s is not zero. Using the notation and transformations of (25), (or direct from (26), (27)), we produce

$$-\frac{dH_1}{dz} = c(s+p)E_1, \quad . \quad . \quad . \quad . \quad (39)$$

$$-\frac{dE_1}{dz} = \mu(-s+p)H_1; \quad . \quad . \quad . \quad (40)$$

from which

$$v^2 \frac{d^2H_1}{dz^2} = (p^2 - s^2)H_1, \quad . \quad . \quad . \quad . \quad (41)$$

with the same equation for E_1.

The complete solution may be thus described. Let, at time $t=0$, there be $H=H_0$ through the small distance a at the origin. This immediately splits into two plane waves of half the amplitude, which travel to right and left respectively at speed v, attenuating as they progress, so that at time t later, when they are at distances $\pm vt$ from the origin, their amplitudes equal

$$\tfrac{1}{2} H_0 \epsilon^{-qt}, \quad . \quad . \quad . \quad . \quad . \quad . \quad . \quad . \quad (42)$$

with corresponding E's, viz.

$$\tfrac{1}{2}\mu v H_0 \epsilon^{-qt} \text{ and } -\tfrac{1}{2}\mu v H_0 \epsilon^{-qt}, \quad . \quad . \quad . \quad (43)$$

on the right and left sides respectively. These extend through the distance a. Between them is a diffused disturbance, given by

$$H = \epsilon^{-qt} \frac{H_0 a}{2v}\left(s + \frac{d}{dt}\right) J_0 \left\{\frac{s}{v}(z^2 - v^2t^2)^{\frac{1}{2}}\right\}, \quad . \quad . \quad . \quad (44)$$

$$E = \epsilon^{-qt} \frac{H_0 a}{2cv}\left(-\frac{d}{dz}\right) J_0 \left\{\frac{s}{v}(z^2 - v^2t^2)^{\frac{1}{2}}\right\}, \quad . \quad . \quad . \quad (45)$$

in which $v^2t^2 > z^2$.

In a similar manner, suppose initially $E=E_0$ through distance a at the origin. Then, at time t later, we have two plane strata of depth a at distance vt to right and left respectively, in which

$$E = \tfrac{1}{2}E_0 \epsilon^{-qt} = \pm \mu v H, \quad . \quad . \quad . \quad . \quad . \quad (46)$$

the + sign to be used in the right-hand stratum, the — in the left. And, between them, the diffused disturbance given by

$$E=\epsilon^{-qt}\frac{E_0 a}{2v}\left(-s+\frac{d}{dt}\right)J_0\left\{\frac{s}{v}(z^2-v^2t^2)^{\frac{1}{2}}\right\}, \quad . \quad . \quad (47)$$

$$H=\epsilon^{-qt}\frac{E_0 a}{2\mu v}\left(-\frac{d}{dz}\right)J_0\left\{\frac{s}{v}(z^2-v^2t^2)^{\frac{1}{2}}\right\}. \quad . \quad . \quad . \quad (48)$$

Knowing thus the effects due to initial elements of E_0 and H_0, we have only to integrate with respect to z to find the solutions due to any arbitrary initial distributions. I forbear from giving a detailed demonstration, leaving the satisfaction of the proper conditions to be the proof of (42) to (48); since, although they were very laboriously worked out by myself, yet, as mathematical solutions, are more likely to have been given before in some other physical problem than to be new.

Another way of viewing the matter is to start with $s=0$, and then examine the effect of introducing s, either + or −. Let an isolated plane disturbance of small depth be travelling along in the positive direction undistorted at speed v. We have $E=\mu v H$ in it. Now suddenly increase k, making s positive. The disturbance still keeps moving on at the same speed, but is attenuated with greater rapidity. At the same time it leaves a tail behind it, the tip of which travels out the other way at speed v, so that at time t, after commencement of the tailing, the whole disturbance extends through the distance $2vt$. In this tail H is of the same sign as in the head, and its integral amount is such that it exactly accounts for the extra-attenuation suffered by H in the head. On the other hand, E in the tail is of the opposite sign to E in the head; so that the integral amount of E in head and tail decreases faster. As a special case, let, in the first place, there be no conductivity, $k=0$ and $g=0$. Then, keeping g still zero, the effect of introducing k is to cause the above-described effect, except that as there was no attenuation at first, the attenuation later is entirely due to k, whilst the line-integral of H along the tail, or

$$\int H dz,$$

including H in the head, remains constant. This is the persistence of momentum.

If, on the other hand, we introduce g, the statements made regarding H are now true as regards E, and conversely. The tail is of a different nature, E being of same sign in the tail as in the head, and H of the opposite sign. Hence, of course, when we have both k and g of the right amounts, there is no tailing. This subject is, however, far better studied in the telegraphic application, owing to the physical reality then existent, than in the present problem, and also then by elementary methods*.

8. Owing to the presence of d/dz in (45) and (47) we are enabled to give some integral solutions in a finite form. Thus, let $H=H_0$ constant and $E=0$ initially on the whole of the negative side of the origin, with no E or H on the positive side. The E at time t later is got by integrating (45), giving

$$E=\frac{H_0}{2cv}J_0\left\{\frac{s}{v}(z^2-v^2t^2)^{\frac{1}{2}}\right\}\epsilon^{-qt}, \quad . \quad . \quad . \quad (49)$$

which holds between the limits $z=\pm vt$, there being no disturbance beyond, except the H_0 on the left side. When $g=0$ and z/vt is small, it reduces to

$$E=\frac{H_0}{4\pi}\left(\frac{\mu}{kt}\right)^{\frac{1}{2}}\epsilon^{-\pi k\mu z^2/t}. \quad . \quad . \quad . \quad . \quad (50)$$

This is the pure diffusion-solution, suitable for good conductors.

If initially $E=E_0$, constant, on the left side of the origin, and zero on the right side, then at time t the H due to it is, by (48),

$$H=\frac{E_0}{2\mu v}J_0\left\{\frac{s}{v}(z^2-v^2t^2)^{\frac{1}{2}}\right\}\epsilon^{-qt}. \quad . \quad . \quad . \quad . \quad (51)$$

The result of taking $c=0$, $g=0$, in this formula is zero, as we may see by observing that c in (49) becomes μ in (51). It is of course obvious that, as the given initial electric field has no energy if $c=0$, it can produce no effect later.

The H solution corresponding to (49) cannot be finitely expressed. It is

$$H=\tfrac{1}{2}H_0\epsilon^{-qt}\left[1+\int_z^{vt}\frac{dx}{v}\left(s+\frac{d}{dt}\right)J_0\left\{\frac{s}{v}(x^2-v^2t^2)^{\frac{1}{2}}\right\}dx\right],$$

which, integrated, gives

$$H=\tfrac{1}{2}H_0\epsilon^{-qt}\Big[\epsilon^{st}-\frac{sz}{v}(J_0-iJ_1)+\frac{1}{\lfloor 3}\left(\frac{sz}{v}\right)^3\frac{1}{st}(-iJ_1-J_2)$$

$$-\frac{1.3}{\lfloor 5}\left(\frac{sz}{v}\right)^5\frac{1}{s^2t^2}(-J_2+iJ_3)+\frac{1.3.5}{\lfloor 7}\left(\frac{sz}{v}\right)^7\frac{1}{s^3t^3}(iJ_3+J_4)+\ldots\Big], \quad (52)$$

* "Electro-magnetic Induction and its Propagation," Electrician, Arts. xlii. to l. (1887).

where all the J's operate on $st\sqrt{-1}$; thus, *e. g.* (Bessel's),

$$J_3 = J_3(st\sqrt{-1}).$$

But a much better form than (52), suitable for calculating the shape of the wave speedily, especially at its start, may be got by arranging in powers of $z-vt$, thus

$$H = \tfrac{1}{2}H_0\epsilon^{-qt}\left\{1 + stf_1\left(1-\frac{z}{vt}\right) + \frac{s^2t^2}{\underline{|2}}f_2\left(1-\frac{z}{vt}\right)^2 + \frac{s^3t^3}{\underline{|3}}f_3\left(1-\frac{z}{vt}\right)^3 + \ldots\right\}, \quad . \quad (53)$$

true when $z < vt$, where f_1, f_2, &c. are functions of t only, of which the first five are given by

$$f_1 = 1 + \frac{st}{2},$$

$$f_2 = \frac{st}{2}\left(1+\frac{st}{4}\right),$$

$$f_3 = -\tfrac{1}{2}\left(1+\frac{st}{4}\right) + \frac{s^2t^2}{2.4}\left(1+\frac{st}{6}\right),$$

$$f_4 = -\tfrac{3}{4}\frac{st}{2}\left(1+\frac{st}{6}\right) + \frac{s^3t^3}{2.4.6}\left(1+\frac{st}{8}\right),$$

$$f_5 = \tfrac{3}{8}\left(1+\frac{st}{6}\right) - \frac{s^2t^2}{2.4}\left(1+\frac{st}{8}\right) + \frac{s^4t^4}{2.4.6.8}\left(1+\frac{st}{10}\right).$$

At the origin, H is given by

$$H = \tfrac{1}{2}H_0\epsilon^{-2q_2t}, \quad . \quad . \quad . \quad . \quad . \quad . \quad . \quad (54)$$

and is therefore permanently $\frac{1}{2}H_0$ when $g=0$. At the front of the wave, where $z=vt$,

$$H = \tfrac{1}{2}H_0\epsilon^{-qt}. \quad . \quad . \quad . \quad . \quad . \quad . \quad . \quad . \quad (55)$$

Now, to represent the E solution corresponding to (51), we have only to turn H_0 to E_0 in (53), and change the sign of s throughout, *i. e.* explicit, and in the f's. Similarly in (52). Thus, at the origin,

$$E = \tfrac{1}{2}E_0\epsilon^{-2q_1t}, \quad . \quad . \quad . \quad . \quad . \quad . \quad . \quad . \quad (56)$$

and at the front of the wave

$$E = \tfrac{1}{2}E_0\epsilon^{-qt}. \quad . \quad . \quad . \quad . \quad . \quad . \quad . \quad . \quad (57)$$

9. Again, let $H=\frac{1}{2}H_0$ on the left side, and $H=-\frac{1}{2}H_0$ on the right side of the origin, initially. The E that results from each of them is the same, and is half that of (49); so that (49) still expresses the E solution. This case corresponds to an initial electric current of surface-density $H_0/4\pi$ on the

$z=0$ plane, with the full magnetic field to correspond, and from it immediately follows the E solution due to any initial distribution of electric current in plane layers.

Owing to H being permanently $\frac{1}{2}H_0$ at the origin in the case (49), (54), when $g=0$, we may state the problem thus: An infinite conducting dielectric with a plane boundary is initially free from magnetic induction, and its boundary suddenly receives the magnetic force $H_0=$constant. At time t later (49) and (52) or (53) give the state of the conductor at distance $z<vt$ from the boundary. In a good conductor the attenuation at the front of the wave is so enormous that the diffusion solution (50) applies practically. It is only in bad conductors that the more complete form is required.

10. *Effect of Impressed Force.*—We can show that the initial effect of impressed force is the same as if the dielectric were nonconducting. In equations (23), (24), let $p=ni$, where $n2\pi=$ frequency of alternations, supposing **e** to alternate rapidly. By increasing n we can make the second terms on the right sides be as great multiples of the first terms as we please, so that in the limit we have results independent of k and g, in this respect, that as the frequency is raised infinitely, the true solutions tend to be infinitely nearly represented by simplified forms, in which k and g play the part of small quantities. An inspection of the sinusoidal solution for plane waves shows that **E** and **H** get into the same phase, and that k and g merely present themselves in the exponents of factors representing attenuation of amplitude as the waves pass away from the seat of vorticity of impressed force.

Consequently, in the plane problem, the initial effect of an abrupt discontinuity in e, say $e=$constant on the left, and zero on the right side of the plane through the origin, is to produce

$$H=-e/2\mu v \quad . \quad . \quad . \quad . \quad . \quad . \quad (58)$$

all over the plane of vorticity; and

$$E=\mp\tfrac{1}{2}e \quad . \quad . \quad . \quad . \quad . \quad . \quad . \quad (59)$$

on its left and right sides respectively. We may regard the plane as continuously emitting these disturbances to right and left at speed v so long as the impressed force is in operation, but their subsequent history can only be fully represented by the tail formulæ already given.

Irrespective of the finite curvature of a surface, any element thereof may be regarded as plane. Therefore every element of a sheet of vortex lines of impressed force acts in the way just described as being true of the elements of an infinite plane sheet. But it is only in comparatively simple cases, of which I shall give examples later, that the subsequent course of events does not so greatly complicate matters as to render it impossible to go into details after the first moment. On first starting the sheet, it becomes a sheet of magnetic induction, whose lines coincide with the vortex lines of impressed force. If f be the measure of the vorticity per unit area, $f/2\mu v$ is the intensity of the magnetic force. In the imaginary good conductor of *no* permittivity, this is zero, owing to v being then assumed to be infinite.

Notice that whilst the vorticity of $\mathbf{e}$ produces magnetic induction, that of $\mathbf{h}$ produces electric displacement, and whilst in the former case $\mathbf{E}$ is made discontinuous at a plane of finite vorticity, in the latter case it is $\mathbf{H}$ that is initially discontinuous.

11. *True Nature of Diffusion in Conductors.*—The process of diffusion of magnetic induction in conductors appears to be fundamentally one of repeated internal reflexions with partial transmission. Thus, let a plane wave $E_1=\mu v H_1$ moving in a nonconducting dielectric strike flush an exceedingly thin sheet of metal. Let $E_2=\mu v H_2$ be the transmitted wave in the dielectric on the other side, and $E_3=-\mu v H_3$ be the reflected wave. At the sheet we have

$$E_1+E_3=E_2, \qquad (60)$$

$$H_1+H_3=H_2+4\pi k_1 z E_2, \qquad (61)$$

if k_1 be the conductivity of the sheet of thickness z. Therefore

$$\frac{E_2}{E_1}=\frac{H_2}{H_1}=\frac{E_1+E_3}{E_1}=\frac{1}{1+2\pi\mu k_1 z v}. \qquad (62)$$

H is reflected positively and E negatively. A perfect conducting barrier is a perfect reflector, it doubles the magnetic force and destroys the electric force on the side containing the incident wave, and transmits nothing.

Take $k_1=(1600)^{-1}$ for copper, and $\mu v=3\times 10^{10}$ centim.

Then we see that to attenuate the incident wave H_1 to $\frac{1}{2}H_1$ by transmission through the plate, requires

$$z=(2\pi\mu k_1 v)^{-1}=\frac{8}{3\pi}\times 10^{-8} \text{ centim.}, \qquad (63)$$

which is a very small fraction of the wave-length of visible light. The H disturbance is made $\frac{3}{2}H_1$, the E reduced to $\frac{1}{2}E_1$ on the transmission side. There is, however, persistence of H, although there is dissipation of E. To produce dissipation of H with persistence of E requires the plate to be a magnetic, not an electric conductor.

Now, imagine an immense number of such plates to be

packed closely together, with dielectric between them, forming a composite dielectric conductor, and let the outermost sheet be struck flush by a plane wave as above. The first sheet transmits $\frac{1}{2}H_1$, the second $\frac{1}{4}H_1$, the third $\frac{1}{8}H_1$, and so on. This refers to the front of the wave, going into the composite conductor at speed v. It is only necessary to go a very short distance to attenuate the front of the wave to nothing; the immense speed of propagation does not result in producing any sensible immediate effect at a distance, which comes on quite slowly as the complex result of all the internal reflexions and transmissions between and at the sheets. Observe that there is an initial accumulation of H, so to speak, at the boundary of the conductor, due to the reflexion. [Example: the current-density may be greater at the outermost layer* of a round wire when the current is started in it than the final value, and the total current in the wire increases faster than if it were constrained to be uniformly distributed.]

Thus a good conductor may have very considerable permittivity, much greater than that of air, and yet show no signs of it, on account of the extraordinary attenuation produced by the conductivity. Now this is rather important from the theoretical point of view. It is commonly assumed that good conductors, *e.g.* metals, are not dielectrics at all. This makes the speed of propagation of disturbances through them infinitely great. Such a hypothesis, however, should have no place in a rational theory, professing to represent transmission in time by stresses in a medium occupying the space between molecules of gross matter. But by admitting that not only bad conductors, but all conductors, are also dielectrics, we do away with the absurdity of infinitely rapid action through infinite distances in no time at all, and make the method of propagation, although it practically differs so greatly from that in a nonconducting dielectric, be yet fundamentally the same, with its characteristic features masked by repeated internal reflexions with loss of energy. We need not take any account of the electric displacement in actual reckonings of the magnitude of the effects which can be observed in the case of good conductors, but it is surely a mistake to overlook it when it is the nature of the actions involved that is in question. (See Note C, p. 153).

Why conductors act as reflectors is quite another question, which can only be answered speculatively. If molecules are perfect conductors, they are perfect reflectors, and if they were packed quite closely, we should nearly have a perfect conductor in mass, impenetrable by magnetic induction; and

* "On the S I. of Wires," Part I. Phil. Mag. August 1886.

we know that cooling a metal and packing the molecules closer does increase its conductivity. But as they do not form a compact mass in any substance, they must always allow a partial transmission of electromagnetic waves in the intervening dielectric medium, and this would lead to the diffusion method of propagation. We do not, however, account in this way for the dissipation of energy, which requires some special hypothesis.

The diffusion of heat, too, which is, in Fourier's theory, done by instantaneous action to infinite distances, cannot be physically true, however insignificant may be the numerical departures from the truth. What can it be but a process of radiation, profoundly modified by the molecules of the body, but still only transmissible at a finite speed? The very remarkable fact that the more easily penetrable a body is to magnetic induction the less easily it conducts heat, in general, is at present a great difficulty in the way, though it may perhaps turn out to be an illustration of electromagnetic principles eventually.

12. *Infinite Series of Reflected Waves. Remarkable Identities. Realized Example.*—When, in a plane-wave problem, we confine ourselves to the region between two parallel planes, we can express our solutions in Fourier series, constructed so as to harmonize with the boundary conditions which represent the effect of the whole of the ignored regions beyond the boundaries in modifying the phenomena occurring within the limited region. Now the effect of the boundaries is usually to produce reflected waves. Hence a solution in Fourier series must usually be decomposable into an infinite series of separate solutions, coming into existence one after the other in time if the speed v be finite, or all in operation at once from the first moment if the speed be made infinite (as in pure diffusion). If the boundary conditions be of a simple nature, this decomposition can sometimes be easily explicitly represented, indicating remarkable identities, of which the following investigation leads to one. We may either take the case of plane waves in a conducting dielectric bounded by infinitely conductive planes, making $E=0$ the boundary condition; or, similarly, by infinitely inductive planes producing $H=0$ at them. But the most practical way, and the most easily followed, is to put a pair of parallel wires in the dielectric, and produce a real problem relating to a telegraph-circuit.

Let A and B be its terminations at $z=0$ and $z=l$ respectively. Let them be short-circuited, producing the terminal conditions $V=0$ at A and B in the absence of impressed force at either place. Now, the circuit being free from charge and current initially, insert a steady impressed force e_0 at A. Required the effect, both in Fourier series and in detail, showing the whole history of the phenomena that result.

Equations (36) and (37) are the fundamental connexions of V and C at any distance z from A. Let R, L, K, S be the resistance, inductance, leakage-conductance, and permittance per unit length of circuit, and

$$s_1 = R/2L, \quad s_2 = K/2S, \quad q = s_1 + s_2, \quad s_0 = s_1 - s_2, \quad . \quad (64)$$

$$\lambda = (m^2v^2 - s_0^2)^{\frac{1}{2}}. \quad . \quad . \quad . \quad . \quad (65)$$

It may be easily shown, by the use of the resistance operator, or by testing satisfaction of conditions, that the required solutions are

$$V = V_0 - \frac{2e_0}{l}\Sigma\frac{m \sin mz}{RK + m^2}\epsilon^{-qt}\left[\cos\lambda t + \frac{q}{\lambda}\sin\lambda t\right], \quad . \quad . \quad . \quad (66)$$

$$C = C_0 - \frac{e_0}{Rl}\epsilon^{-2s_1t} - \frac{2e_0}{Rl}\Sigma\frac{\cos mz}{m^2 + RK}\epsilon^{-qt}\left[RK\left(\cos - \frac{s_0}{\lambda}\sin\right)\lambda t - 2s_1m^2\frac{\sin\lambda t}{\lambda}\right], \quad . \quad . \quad (67)$$

where $m = j\pi/l$, and j includes all integers from 1 to ∞; whilst V_0 and C_0 represent the final steady V and C, which are

$$V_0 = e_0\left(\cos m_0z - \frac{\sin m_0z}{\tan m_0l}\right), \quad . \quad . \quad . \quad . \quad (68)$$

$$C_0 = \frac{m_0e_0}{R}\left(\sin m_0z + \frac{\cos m_0z}{\tan m_0l}\right), \quad . \quad . \quad . \quad (69)$$

where $m_0^2 = -RK$.

Now if the circuit were infinitely long both ways and were charged initially to potential-difference $2e_0$ on the whole of the negative side of A, with no charge on the positive side, and no current anywhere, the resulting current at time t later at distance z from A would be

$$C_1 = \frac{e_0}{Lv}\epsilon^{-qt}\,J_0\left\{\frac{s_0}{v}(z^2 - v^2t^2)^{\frac{1}{2}}\right\}, \quad . \quad . \quad (70)$$

by §§ 7 and 8; and if, further, K = 0, V at A would be permanently e_0, which is what it is in (66). Hence the C solution

(67) can be finitely decomposed into separate solutions of the form (70) in the case of perfect insulation, when (67) takes the form

$$C=\frac{e_0}{Rl}(1-\epsilon^{-2qt})+\frac{2e_0}{Ll}\epsilon^{-qt}\Sigma\cos mz\frac{2q}{\lambda}\sin\lambda t, \quad . \quad (71)$$

where $q=s_1=s_0$, by the vanishing of s_2 in (64).

Therefore (70) represents the real meaning of (71) from $t=0$ to l/v, provided $vt>z$. But on arrival of the wave C_1 at B, V becomes zero, and C doubled by the reflected wave that then commences to travel from B to A. This wave may be imagined to start when $t=0$ from a point distant l beyond B, and be the precise negative of the first wave as regards V and the same as regards C. Thus

$$C_2=\frac{e_0}{Lv}\epsilon^{-qt}J_0\left\{\frac{q}{v}[(2l-z)^2-v^2t^2]^{\frac{1}{2}}\right\}, \quad . \quad . \quad (72)$$

expresses the second wave, starting from B when $t=l/v$, and reaching A when $t=2l/v$. The sum of C_1 and C_2 now expresses (71), where the waves coexist, and C_1 alone expresses (71) in the remainder of the circuit.

The reflected wave arising when this second wave reaches A may be imagined to start when $t=0$ from a point distant $2l$ from A on its negative side, and be a precise copy of the first wave. Thus

$$C_3=\frac{e_0}{Lv}\epsilon^{-qt}J_0\left\{\frac{q}{v}[(2l+z)^2-v^2t^2]^{\frac{1}{2}}\right\}, \quad . \quad . \quad (73)$$

expresses the third wave; and now (71) means $C_1+C_2+C_3$ in those parts of the circuit reached by C_3 and C_1+C_2 in the remainder.

The fourth wave is, similarly,

$$C_4=\frac{e_0}{Lv}\epsilon^{-qt}J_0\left\{\frac{q}{v}[(4l-z)^2-v^2t^2]^{\frac{1}{2}}\right\}, \quad . \quad . \quad (74)$$

starting from B when $t=3l/v$, and reaching A when $t=4l/v$. And so on, *ad inf.**

* It is not to be expected that in a real telegraph-circuit the successive waves have abrupt fronts, as in the text. There are causes in operation to prevent this, and round off the abruptness. The equations connecting

If we take $L=0$ in this problem, we make $v=\infty$, and bring the whole of the waves into operation immediately. (70) becomes

$$C_1=e_0\left(\frac{S}{\pi Rt}\right)^{\frac{1}{2}}\epsilon^{-RSz^2/4t}; \quad . \quad . \quad . \quad . \quad (75)$$

and similarly for C_2, C_3, &c. In this simplified form the identity is that obtained by Sir W. Thomson* in connexion with his theory of the submarine cable; also discussed by A. Cayley* and J. W. L. Glaisher†.

In order to similarly represent the history of the establishment of V_0, we require to use the series (53) or some equivalent. In other respects there is no difference.

Whilst it is impossible not to admire the capacity possessed by solutions in Fourier series to compactly sum up the effect of an infinite series of successive solutions, it is greatly to be regretted that the Fourier solutions themselves should be of such difficult interpretation. Perhaps there will be discovered some practical way of analyzing them into easily interpretable forms.

Some special cases of (66), (67) are worthy of notice. Thus V is established in the same way when $R=0$ as when $K=0$, provided the value of K/S in the first case be the same as that of R/L in the second. Calling this value $2q$, we have in both cases

$$V=e_0\left(1-\frac{z}{l}\right)-\frac{2e_0}{l}\epsilon^{-qt}\Sigma\frac{\sin mz}{m}\left(\cos\lambda t+\frac{q}{\lambda}\sin\lambda t\right). \quad (76)$$

But the current is established in quite different manners. When it is K that is zero, (71) is the solution; but if R vanish instead, then (67) gives

$$C=\frac{e_0t}{Ll}+\frac{e_0Kl}{2}\left(1-\frac{z}{l}\right)^2-\frac{2e_0K}{l}\epsilon^{-qt}\Sigma\frac{\cos mz}{m^2}\left\{\cos\lambda t-\left(\frac{m^2v^2}{2q}-q\right)\frac{\sin\lambda t}{\lambda}\right\}. \quad (77)$$

V and C express the first approximation to a complete theory. Thus the wires are assumed to be instantaneously penetrated by the magnetic induction as a wave passes over their surfaces, as if the conductors were infinitely thin sheets of the same resistance. It is only a very partial remedy to divide a wire into several thinner wires, unless we at the same time widely separate them. If kept quite close it would, with copper, be no remedy at all.

* Math. and Physical Papers, vol. ii. Art. lxxii.; with Note by A. Cayley.

† Phil. Mag. June 1874.

C now mounts up infinitely. But the leakage-current, which is KV, becomes steady, as (76) shows.

In connexion with this subject I should remark that the non-distortional circuit produced by taking $R/L=K/S$ is of immense assistance, as its properties can be investigated in full detail by elementary methods, and are most instructive in respect to the distortional circuits in question above *.

13. *Modifications made by Terminal Apparatus. Certain cases easily brought to full realization.*—Suppose that the terminal conditions in the preceding are $V=-Z_0C$ and $V=Z_1C$, Z_0 and Z_1 being the "resistance operators" of terminal apparatus at A and B respectively. In a certain class of cases the determinantal equation so simplifies as to render full realization possible in an elementary manner. Thus, the resistance-operator of the circuit, reckoned at A, is †

$$\phi=Z_0+\frac{(R+Lp)l(\tan ml)/ml+Z_1}{1+(K+Sp)lZ_1(\tan ml)/ml}, \quad . \quad (78)$$

where

$$m^2=-(R+Lp)(K+Sp). \quad . \quad . \quad . \quad . \quad . \quad . \quad (79)$$

That is, $e=\phi C$ is the linear differential equation of the current at A. Now, to illustrate the reductions obviously possible, let $Z_0=0$, and

$$Z_1=n_1l(R+Lp). \quad . \quad . \quad . \quad . \quad . \quad . \quad (80)$$

This makes the apparatus at B a coil whose time-constant is L/R, and reduces ϕ to

$$\phi=(R+Lp)l\left(\frac{\tan ml}{ml}+n_1\right)\left\{1-m^2n_1l^2\frac{\tan ml}{ml}\right\}^{-1}, \quad . \quad (81)$$

so that the roots of $\phi=0$ are given by

$$R+Lp \qquad =0. \quad . \quad . \quad . \quad . \quad . \quad . \quad (82)$$

$$\tan ml+mln_1=0; \quad . \quad . \quad . \quad . \quad . \quad . \quad (83)$$

i. e. a solitary root $p=-R/L$ and the roots of (83), which is an elementary well-known form of determinantal equation.

The complete solution due to the insertion of the steady impressed force e_0 at A will be given by‡

* "Electromagnetic Induction and its Propagation," Arts. xl. to l.

† "On the Self-Induction of Wires," Part IV.

‡ *Ib.* Parts III. and IV. Phil. Mag. Oct. and Nov. 1886; or "On Resistance and Conductance Operators," Phil. Mag. Dec. 1887, § 17, p. 500.

$$V = V_0 + \Sigma e_0 u \epsilon^{pt} \div \left(p \frac{d\phi}{dp}\right), \quad . \quad . \quad . \quad . \quad (84)$$

$$C = C_0 + \Sigma e_0 w \epsilon^{pt} \div \left(p \frac{d\phi}{dp}\right), \quad . \quad . \quad . \quad . \quad (85)$$

where the summations range over all the p roots of $\phi = 0$, subject to (79); whilst u and w are the V and C functions in a normal system expressed by

$$w = \cos mz, \quad u = m \sin mz \div (K + Sp); \quad . \quad . \quad (86)$$

and V_0, C_0 are the final steady V and C. In the case of the solitary root (82) we shall find

$$-p \frac{d\phi}{dp} = Rl(1 + n_1), \quad . \quad . \quad . \quad . \quad . \quad . \quad . \quad . \quad . \quad (87)$$

but for all the rest

$$-p \frac{d\phi}{dp} = \frac{l}{2(K + Sp)} \frac{dm^2}{dp} (1 + n_1 \cos^2 ml). \quad . \quad . \quad (88)$$

Realizing (84), (85) by pairing terms belonging to the two p's associated with one m^2 through (79), we shall find that (66), (67) express the solutions, provided we make these simple changes:—Divide the general term in both the summations by

$$1 + n_1 \cos^2 ml,$$

and the term following C_0 outside the summation in (67) by $(1 + n_1)$. Of course the m's have now different values, as per (83), and V_0, C_0 are different.

14. There are several other cases in which similar reductions are possible. Thus, we may have

$$Z_0 = n_0(R + Lp) + n_0'(K + Sp)^{-1},$$

$$Z_1 = n_1(R + Lp) + n_1'(K + Sp)^{-1},$$

simultaneously, n_0, n_0', n_1, n_1' being any lengths. That is, apparatus at either end consisting of a coil and a condenser in sequence, the time-constant of the coil being L/R and that of the condenser S/K. Or, the condenser may be in parallel with the coil. In general we have, as an alternative form of $\phi = 0$, equation (78),

$$\frac{\tan ml}{ml} = -\frac{(Z_0 + Z_1)\{(R + Lp)l\}^{-1}}{1 - m^2 l^2 Z_0 Z_1 \{(R + Lp)l\}^{-2}}; \quad . \quad . \quad (89)$$

from which we see that when

$$\frac{Z_0}{(R+Lp)l} \quad \text{and} \quad \frac{Z_1}{(R+Lp)l}$$

are functions of ml, equation (89) finds the value of m^2 immediately, *i. e.* not indirectly as functions of p. In all such cases, therefore, we may advantageously have the general solutions (80), (81) put into the realized form. They are

$$V=V_0-\frac{2e_0}{l}\Sigma\frac{(\sin mz+\tan\theta\cos mz)m\epsilon^{-qt}(\cos+q\lambda^{-1}\sin)\lambda t}{\sec^2\theta(m^2+RK)\left(1-\cos^2 ml\,\frac{d}{d(ml)}\tan ml\right)}, \quad . \quad (90)$$

$$C=C_0-\frac{2e_0}{l}\Sigma\frac{(\cos mz-\tan\theta\sin mz)\epsilon^{-qt}K\{\cos-(2s_2\lambda)^{-1}(\lambda^2+qs_0)\sin\}\lambda t}{\text{same denominator}}, \quad (91)$$

where q, λ, s_0, s_2 are as in (64), (65). The differentiation shown in the denominator is to be performed upon the function of ml to which $\tan ml$ is equated in (89) after reduction to the form of such a function in the way explained; and θ depends upon Z_0 thus,

$$\left.\begin{aligned}\tan\theta&=-m^{-1}(K+Sp)Z_0,\\ \sec^2\theta&=1+m^{-2}Z_0^2(K+Sp)^2,\end{aligned}\right\}, \quad . \quad . \quad . \quad (92)$$

which are also functions of ml. It should be remarked that the terms depending upon solitary roots, occurring in the case $m^2=0$, are not represented in (90), (91). They must be carefully attended to when they occur.

Note A.

An electromagnetic theory of light becomes a necessity, the moment one realizes that it is the same medium that transmits electromagnetic disturbances and those concerned in common radiation. Hence *the* electromagnetic theory of Maxwell, the essential part of which is that the vibrations of light are really electromagnetic vibrations (whatever they may be), and which is an undulatory theory, seems to possess far greater intrinsic probability than *the* undulatory theory, because that is not an electromagnetic theory. Adopting, then, Maxwell's notion, we see that the only difference between the waves in telephony (apart from the distortion and dissipation due to resistance) and light-waves is in the wave-length; and the fact that the speed, as calculated by electromagnetic data, is the same as that of light, furnishes a powerful argument in favour of the extreme relative simplicity of constitution of the æther, as compared with common matter in bulk. There is observational reason to believe that the sun sometimes causes magnetic disturbances here of the ordinary kind. It is impossible to attribute this to any amount of increased activity of emission of the sun so long as we only think of common radiation. But, bearing in mind the long waves of electromagnetism, and the constant speed, we see that disturbances from the sun may be hundreds or thousands of miles long of one kind (*i. e.* without alternation), and such waves, in passing the earth, would cause magnetic "storms," by inducing currents in the earth's crust and in telegraph-wires. Since common radiation is ascribed to molecules, we must ascribe the great disturbances to movements of large masses of matter.

There is nothing in the abstract electromagnetic theory to indicate whether the electric or the magnetic force is in the plane of polarization, or rather, surface of polarization. But by taking a concrete example, as the reflexion of light at the boundary of transparent dielectrics, we get Fresnel's formula for the ratio of reflected to incident wave, on the assumption that his "displacement" coincides with the electric displacement; and so prove that it is the magnetic flux that is in the plane of polarization.

Note B.

I give these numerical examples:—

Take a circuit 100 kilom. long, of 4 ohms and $\frac{1}{4}$ microf. per kilom. and *no* inductance in the first place, and also no leakage in any case. Short-circuit at beginning A and end B. Introduce at A a sinusoidal impressed force, and calculate the amplitude of the current at B by the electrostatic theory. Let the ratio of the full steady current to the amplitude of the sinusoidal current be ρ, and let the frequency range through 4 octaves, from $n=1250$ to $n=20{,}000$; the frequency being $n/2\pi$. The values of ρ are

1·723, 3·431, 10·49, 58·87, 778.

It is barely credible that any kind of speaking would be possible, owing to the extraordinarily rapid increase of attenuation with the frequency. Little more than murmuring would result.

Now make $L=2\frac{1}{2}$ (very low indeed), L being inductance per centim. Calculate by the combined electrostatic and magnetic formulæ. The corresponding figures are

1·567, 2·649, 5·587, 10·496, 16·607.

The change is marvellous. It is only by the preservation of the currents of great frequency that good articulation is possible, and we see that even a very little self-induction immensely improves matters. There is no "dominant" frequency in telephony. What should be aimed at is to get currents of any frequency reproduced at B in their proper proportions, attenuated to the same extent.

Change L to 5. Results:—

1·437, 2·251, 3·176, 4·169, 4·670.

Good telephony is now possible, though much distortion remains.

Increase L to 10. Results:—

1·235, 1·510, 1·729, 1·825, 1·854.

This is first class, showing approximation towards a non-distortional circuit. Now this is all done by the self-induction carrying forward the waves undistorted (relatively) and also with much less attenuation.

I should add that I attach no importance to the above figures in point of exactness. The theory is only a first approximation. In order to emphasize the part played by self-induction, I have stated that by sufficiently increasing it (without other change, if this could be possible) we could make the amplitude of current at the end of an Atlantic cable greater than the steady current (by the *quasi*-resonance).

Note C.

In Sir W. Thomson's article on the "Velocity of Electricity" (Nichols's *Cyclopædia*, 2nd edition, 1860, and Art. lxxxi. of 'Mathematical and Physical Papers,' vol. ii.) is an account of the chief results published up to that date relating to the "velocity" of transmission of electricity, and a very explicit statement, except in some respects as regards inertia, of the theoretical meaning to be attached to this velocity under different circumstances. This article is also strikingly illustrative of the remarkable contrast between Sir W. Thomson's way of looking at things electrical (at least at that time) and Maxwell's views; or perhaps I should say Maxwell's plainly evident views combined with the views which his followers have extracted from that mine of wealth 'Maxwell,' but which do not lie on the surface. [As charity begins at home, I may perhaps illustrate by a personal example the difference between the patent and the latent, in Maxwell. If I should claim (which I do) to have discovered the true method of establishment of current in a wire—that is, the current starting on its boundary, as the result of the initial dielectric wave outside it, followed by diffusion inwards,—I might be told that it was all "in Maxwell." So it is; but entirely latent. And there are many more things in Maxwell which are not yet discovered.] This difference has been the subject of a most moving appeal from Prof. G. F. Fitzgerald, in 'Nature,' about three years since. There really seemed to be substance in that appeal. For it is only a master-mind that can adequately attack the great constructive problem of the æther, and its true relation to matter; and should there be reason to believe that the master is on the wrong track, the result must be, as Prof. Fitzgerald observed (in effect) disastrous to progress. Now Maxwell's theory and methods have stood the test of time, and showed themselves to be eminently rational and developable.

It is not, however, with the general question that we are here concerned, but with the different kinds of "velocity of electricity." As Sir W. Thomson points out, his electrostatic theory, by ignoring electromagnetic induction, leads to infinite speed of electricity through the wire. Interpreted in terms of Maxwell's theory, this speed is not that of electricity through the wire at all, but of the waves through the dielectric, guided by the wire. It results, then, from the assumption $\mu=0$, destroying inertia (not of the electric current, but of the magnetic field), and leaving only forces of elasticity and resistance.

But he also points out another way of getting an infinite speed, when we, in the case of a suspended wire, not of great length, ignore the static charge. This is illustrated by the pushing of incompressible water through an unyielding pipe, constraining the current to be the same in all parts of the circuit. This, in Maxwell's theory, amounts to stopping the elastic displacement in the dielectric, and so making the speed of the wave through it infinite. As, however, the physical actions must be the same, whether a wire be long or short, the assumption being only warrantable for purposes of calculation, I have explained the matter thus. The electromagnetic waves are sent to and fro with such great frequency (owing to the shortness of the line) that only the mean value of the oscillatory V at any part can be perceived, and this is the final value; at the same time, by reason of current in the negative waves being of the same sign as in the positive, the current C mounts up by little jumps, which are, however, packed so closely together as to make a practically continuous rise of current in a smooth curve, which is that given by the electromagnetic theory. This curve is of course practically the same all over the circuit, because of the little jumps being imperceptible.

But in any case this speed is not the speed of electricity through the wire, but through the dielectric outside it. Maxwell remarked that we know nothing of the speed of electricity in a wire supporting current; it may be an inch in an hour, or immensely great. This is on the assumption, apparently, that the electric current in a wire really consists in the transfer of electricity through the wire. I have been forced, to make Maxwell's scheme intelligible to myself, to go further, and add that the electricity may be standing still, which is as much as to say that there is no current, in a literal sense, inside a conductor. [The slipping of electrification over the surface of a wire is quite another thing. That is merely the movement of the wave through the dielectric, guided by the wire. It occurs in a non-distortional circuit, owing to the absence of tailing, in the most plainly evident manner.] In other words, take Maxwell's definition of electric current in terms of magnetic force as a basis, and ignore the imaginary fluid behind it as being a positive hindrance to progress, as soon as one leaves the elementary field of *steady* currents and has to deal with variable states.

The remarks in the text on the subject of the speed of waves in conductors relates to a speed that is not considered in Sir W. Thomson's article. It is the speed of transmission of magnetic disturbances into the wire, in cylindrical waves, which begins at any part of a wire as soon as the primary wave through the dielectric reaches that part. It would be no use trying to make signals through a wire if we had not the outer dielectric to carry the magnetizing and electrizing force to its boundary. The slowness of diffusion in large masses is surprising. Thus a sheet of copper covering the earth, only 1 centim. in thickness, supporting a current whose external field imitates that of the earth, has a time-constant of about a fortnight. If the copper extended to the centre of the earth, the time-constant of the slowest subsiding normal system would be millions of years.

In the article referred to, Sir W. Thomson mentions that Kirchhoff's investigation, introducing electromagnetic induction, led to a velocity of electricity considerably greater than* that of light, which is so far in accordance with Wheatstone's observation. Now it seems to me that we have here a suggestion of a probable

* [*Note by* SIR WILLIAM THOMSON.] In this statement I inadvertently did injustice to Kirchhoff. In the unpublished investigation referred to in the article *Electricity, Velocity of* [Nichols's *Cyclopædia*, second edition, 1860; or my 'Collected Papers,' vol. ii. page 135 (3)], I had found that the ultimate velocity of propagation of electricity in a long insulated wire in air is equal to the number of electrostatic units in the electromagnetic unit; and I had correctly assumed that Kirchhoff's investigation led to the same result. But, owing to the misunderstanding of two electricities or one, referred to in § 317 of my 'Electrostatics and Magnetism,' I imagined Weber's measurement of the number of electrostatic units in the electromagnetic to be $2\times3{\cdot}1\times10^{10}$ centimetres per second, which would give for the ultimate velocity of electricity through a long wire in air twice the velocity of light. In my own investigation, for the submarine cable, I had found the ultimate velocity of electricity to be equal to the number of electrostatic units in the electromagnetic unit divided by $\sqrt{k}$; k denoting the specific inductive capacity of the gutta-percha. But at that time no one in Germany (scarcely any one out of England) believed in Faraday's "specific inductive capacity of a dielectric."

Kirchhoff himself was perfectly clear on the velocity of electricity in a long insulated wire in air. In his original paper, "Ueber die Bewegung der Electricität in Drähten" (Pogg. *Ann.* Bd. c. 1857; see pages 146 and 147 of Kirchhoff's Volume of Collected Papers, Leipzig, 1882), he gives it as $c/\sqrt{2}$, which is what I then called the number of electrostatic units in the electromagnetic unit; and immediately after this he says, "ihr Werth ist der von 41950 Meilen in einer Sekunde, also sehr nahe gleich der Geschwindigkeit des Lichtes im leeren Raume."

Thus clearly to Kirchhoff belongs the priority of the discovery that the velocity of electricity in a wire insulated in air is very approximately equal to the velocity of light.

explanation of why Sir W. Thomson did not introduce self-induction into his theory. There were presumably more ways than one of doing it, as regards the measure of the electric force of induction. When we follow Maxwell's equations, there is but one way of doing it, which is quite definite, and leads to a speed which cannot possibly exceed that of light, since it is the speed $(\mu c)^{-\frac{1}{2}}$ through the dielectric, and cannot be sensibly greater than 3×10^{10} centim., though it may be less. Kirchhoff's result is therefore in conflict with Maxwell's statement that the German methods lead to the same results as his. Besides that, Wheatstone's classical result has not been supported by any later results, which are always less than the speed of light, as is to be expected (even in a non-distortional circuit). But a reference to Wheatstone's paper on the subject will show, first, that there was confessedly a good deal of guesswork; and, next, that the repeated doubling of the wire on itself made the experiment, from a modern point of view, of too complex a theory to be examined in detail, and unsuitable as a test.

From the PHILOSOPHICAL MAGAZINE, for March 1888.

Note on a Paper on Electromagnetic Waves.
By OLIVER HEAVISIDE*.

AN editorial query, the purport of which I did not at first understand, has directed my attention to Prof. J. J. Thomson's paper "On Electrical Oscillations in Cylindrical Conductors" (Proc. Math. Soc. vol. xvii. Nos. 272, 273), a copy of which the author has been so good as to send me. His results, for example, that an iron wire of $\frac{1}{6}$ centim. radius, of inductivity 500, carries a wave of frequency 100 per second about 100,000 miles before attenuating it from 1 to ϵ^{-1}, and similar results, summed up in his conclusion that the carrying-power of an iron wire cable is very much greater than that of a copper one of similar dimensions, are so surprisingly different from my own, deduced from my developed sinusoidal solutions, in the accuracy of which I have perfect confidence (having had occasion last winter to make numerous practical applications of them in connexion with a paper which was to have been read at the S. T. E. and E.), that I felt sure there must be some serious error of a fundamental nature running through his investigations. On examination I find this is the case, being the use of an erroneous boundary condition in the beginning, which wholly vitiates the subsequent results. It is equivalent to assuming that the tangential component of the flux magnetic induction is continuous at the surface of separation of the wire and dielectric, where the inductivity changes value, from a large value to unity, when the wire is of iron. The true conditions are continuity of tangential *force* and of normal *flux*.

As regards my own results, and how increasing the inductance is favourable, the matter really lies almost in a nutshell; thus. In order to reduce the full expression of Maxwell's connexions to a practical working form I make two assumptions. First, that the longitudinal component of current (parallel to the wires) in the dielectric is negligible, in comparison with the total current in the conductors, which makes C one of the variables, C being the current in either conductor; and next, what is equivalent to supposing that the wave-length of disturbances transmitted along the wires is a large multiple of their distance apart. The result is that the equations connecting V and C become

* This Note may be regarded as a continuation of Note B to "Electromagnetic Waves," Phil. Mag. February 1888.

$$-\frac{dV}{dz}=R''C, \quad -\frac{dC}{dz}=KV+S\frac{dV}{dt};$$

S being the permittance and K the conductance of the dielectric per unit length of circuit, whilst R'' is a "resistance-operator," depending upon the conductors, and their mutual position, which, in the sinusoidal state of variation, reduces to

$$R''=R'+L'\frac{d}{dt},$$

where R' and L' are the effective resistance and inductance of the circuit respectively, per unit length, to be calculated entirely upon electromagnetic principles. It follows that the fully developed sinusoidal solution is of precisely the same form as if the resistance and inductance were constants. Disregarding the effect of reflexions, we have

$$V=V_0\epsilon^{-Pz}\sin(nt-Qz),$$

due to $V_0 \sin nt$ impressed at $z=0$; where P and Q are functions of R', L', S, K, and n.

Now if $R'/L'n$ is large, and leakage is negligible (a well-insulated slowly worked submarine cable, and other cases), we have

$$P=Q=(\tfrac{1}{2}RSn)^{\frac{1}{2}},$$

as in the electrostatic theory of Sir W. Thomson. There is at once great attenuation in transit, and also great distortion of arbitrary waves, owing to P and Q varying with n.

But in telephony, n being large, P and Q may have widely different values, because $R'/L'n$ may be quite small, even a fraction. In such case we have no resemblance to the former results. If $R'/L'n$ is small, P and Q approximate to

$$P=\frac{R'}{2L'v'}+\frac{K}{2Sv'}, \quad Q=\frac{n}{v'},$$

where $v'=(L'S)^{-\frac{1}{2}}$. This also requires K/Sn to be small. But it is always very small in telephony.

Now take the case of copper wires of low resistance. L' is practically L_0, the inductance of the dielectric, and v' is practically v, the speed of undissipated waves, or of all elementary disturbances, through the dielectric, whilst R' may be taken to be R, the steady resistance, except in extreme cases. Hence, with perfect insulation,

$$P=\frac{R}{2L_0v}, \quad Q=\frac{n}{v},$$

or the speed of the waves is v, and the attenuating coefficient P is practically independent of the frequency, and is made smaller by reducing the resistance and *increasing* the inductance, *of the dielectric.*

The corresponding current is

$$C = V/L_0 v$$

very nearly, or V and C are nearly in the same phase, like undissipated plane waves. There is very little distortion in transit.

How to increase L_0 is to separate the conductors, if twin wires, or raise the wire higher from the ground, if a single wire with earth-return. It is not, however, to be concluded that L_0 could be increased indefinitely with advantage. If l is the length of the circuit,

$$Rl = 2L_0 v$$

shows the value of L_0 which makes the received current greatest. It is then far greater than is practically wanted, so that the difficulty of increasing L_0 sufficiently is counterbalanced by the non-necessity. The best value of L_0 is, in the case of a long line, out of reach; so that we may say, generally, that increasing the inductance is always of advantage to reduce the attenuation and the distortion.

Now if we introduce leakage, such that $R/L_0 = K/S$, we entirely remove the distortion, not merely when $R/L_0 n$ is small but of any sort of waves. It is, however, at the expense of increased attenuation. The condition of greatest received current, L_0 being variable, is now

$$Rl = L_0 v.$$

We have thus two ways of securing good transmission of electromagnetic waves: one very perfect, for any kind of signals; the other less perfect, and limited to the case of $R/L_0 n$ small, but quite practical. The next step is to secure that the receiving-instrument shall not introduce further distortion by the quasi-resonance that occurs. In the truly non-distortional circuit this can be done by making the resistance of the eceiver to be $L_0 v$ (whatever the length of the line); this causes complete absorption of the arriving waves. In the other case, of R/Ln small, with good insulation, we require the resistance of the receiver to be $L_0 v$ to secure this result approximately. I have also found that this value of the receiver's resistance is exactly the one that (when size of wire in receiver is variable) makes the magnetic force, and therefore the strength of signal, a maximum. Some correction is required on account of the self-induction of the receiver;

but in really good telephones of the best kind, with very small time-constants, it is not great. We see therefore that telephony, so far as the electrical part of the matter is concerned, can be made as nearly perfect as possible on lines of thousands of miles in length. But the distortion that is left, due to imperfect translation of sound-waves into electromagnetic waves at the sending-end, and the reproduction of sound-waves at the receiving-end, is still very great; though, practically, any fairly good telephonic speech is a sufficiently good imitation of the human voice.

There is one other way of increasing the inductance which I have described, viz. in the case of covered wires to use a dielectric impregnated with iron dust. I have proved experimentally that L_0 can be multiplied several times in this way without any increase of resistance; and the figures I have given above (in Note B) prove what a wonderful difference the self-induction makes, even in a cable, if the frequency is great. Hence, if this method could be made practical, it would greatly increase the distance of telephony through cables.

Now, passing to iron wires, the case is entirely different, on account of the great increase in resistance that the substitution of iron for copper of the same size causes, which increases P and the attenuation. Taking for simplicity the very extreme case of such an excessive frequency as to make the formula

$$R' = (\tfrac{1}{2}R\mu n)^{\frac{1}{2}}$$

nearly true, R being the steady and R the actual resistance, we see that increasing either R or μ increases R' and therefore P, because $L'v'$ tends to the value $L_0 v$. Thus the carrying-power of iron is not greatly above, but greatly below that of copper of the same size.

I have, however, pointed out a possible way of utilizing iron (other than that above mentioned), viz. to cover a bundle of fine iron wires with a copper sheath. The sheath is to secure plenty of conductance; the division of the iron to facilitate the penetration of current, and so lower the resistance still more, to the greatest extent, whilst at the same time increasing the inductance. But the theory is difficult, and it is doubtful whether this method is even theoretically legitimate. First class results were obtained by Van Rysselberghe on a 1000-mile circuit in America (2000 miles of wire), using copper-covered steel wire. Here the resistance was very low, on account of the copper, and the inductance considerable, on account of the dielectric alone; so that there is no certain evidence that the iron did any good except by lowering the resistance. But about the advantage of increasing the inductance of the dielectric there can, I think, be no question. It imparts momentum to the waves and carries them on.

In Note B to the first portion of my paper "On Electromagnetic Waves" (Phil. Mag. Feb. 1888), I gave four sets of numerical results showing the influence of increasing the inductance, selecting a cable of large permittance (constant) in order to render the illustrations more forcible. I take this opportunity of stating that the second set of figures relates to the value $L=2\frac{1}{2}$, not 2, of the inductance per centim. The formula used was equation (82), Part II. of my paper "On the Self-Induction of Wires" (Phil. Mag. Sept. 1886, p. 284), which is

$$C_0=2V_0\frac{(Sn)^{\frac{1}{2}}}{(R'^2+L'^2n^2)^{\frac{1}{4}}}(\epsilon^{2Pl}+\epsilon^{-2Pl}-2\cos 2Ql)^{-\frac{1}{2}};$$

where

$$P \text{ or } Q=(\tfrac{1}{2}Sn)^{\frac{1}{2}}\{(R'^2+L'^2n^2)^{\frac{1}{2}}\mp L'n\}^{\frac{1}{2}};$$

where C_0 is the amplitude of current at $z=l$ due to impressed force $V_0 \sin nt$ at $z=0$, with terminal short-circuits. When the circuit is long enough to make ϵ^{-Pl} small, we obtain

$$\rho=\frac{(R'^2+L'^2n^2)^{\frac{1}{4}}}{2Rl(Sn)^{\frac{1}{2}}}\epsilon^{Pl}$$

as the expression for the ratio ρ of the steady current to the amplitude of the sinusoidal current.

The following table is constructed to show the fluctuating manner of variation of the amplitude with the frequency. Drop the accents, and let R/Ln be small. Then, approximately,

$$\rho=\frac{1}{2y}\left(\epsilon^{y}+\epsilon^{-y}-2\cos 2\frac{nl}{v}\right)^{\frac{1}{2}},$$

where

$$y=Rl/Lv,$$

under no restriction as regards the length of the circuit. Now give y a succession of values, and calculate ρ with the cosine taken as -1, 0, and $+1$. Call the results the maximum, mean, and minimum values of ρ.

y.	Min. ρ.	Mean ρ.	Max. ρ.	y.	ρ.	y.	ρ.
$\frac{1}{2}$	·505	1·500	2·063	6	1·678	12	16·81
1	·521	·878	1·128	7	2·365	14	39·3
2	·587	·686	·771	8	3·378	16	93·2
2·065	·594	·685	·766	9	5·000	18	225
3	·710	·748	·784	10	7·420	20	550
4	·907	·924	·940				
5	1·210	1·218	1·226				

It will be seen that when the resistance of the circuit is only a small multiple of, or is of about the same magnitude as Lv (which may be from 300 to 600 ohms in the case of a suspended copper wire), the variation in the value of ρ as the frequency changes through a sufficiently wide range, is great, merely by reason of the reflexions causing reinforcement or reduction of the strength of the received current. The theoretical least value of ρ is $\frac{1}{2}$, when R/Ln is vanishingly small, indicating a doubling of the amplitude of current. But as y increases the range of ρ gets smaller and smaller. After $y=5$ it is negligible.

It is, however, the mean ρ that is of most importance, because the influence of terminal resistances is to lower the range in ρ, and to a variable extent. The value $y=2\cdot065$, or, practically, $Rl=2Lv$, makes the mean ρ a minimum. As I pointed out in the paper before referred to, these fluctuations can only be prejudicial to telephony. In the present Note I have described how to almost entirely destroy them. The principle may be understood thus. Let the circuit be infinitely long first. Then its impedance to an intermediate impressed force alternating with sufficient frequency to make R/Ln small will be $2Lv$, viz. Lv each way. The current and potential-difference produced will be in the same phase, and in moving away from the source of energy they will be similarly attenuated according to the time-factor $\epsilon^{-Rt/2L}$. In order that the circuit, when of finite length, shall still behave as if of infinite length, the constancy of the impedance suggests to us that we should make the terminal apparatus a mere resistance, of amount Lv, by which the waves will be absorbed without reflexion.

That this is correct we may prove by my formula fo the amplitude of received current when there is terminal apparatus, equation (19 *b*), Part V. "On the Self-Induction of Wires" (Phil. Mag. Jan. 1887). It is

$$C_0=2V_0\left[\frac{K^2+S^2n^2}{R'^2+L'^2n^2}\right]^{\frac{1}{4}}\left[G_0G_1\,\epsilon^{2Pl}+H_0H_1\,\epsilon^{-2Pl}\right.$$
$$\left.-2(G_0G_1H_0H_1)^{\frac{1}{2}}\cos 2(Ql+\theta)\right]^{-\frac{1}{2}}.$$

Here C_0 is the amplitude of received current at $z=l$ due to $V_0 \sin nt$ impressed force at $z=0$; R' and L' the effective resistance and inductance per unit length of circuit; K and S the leakage-conductance and permittance per unit-length,

$$P \text{ or } Q=(\tfrac{1}{2})^{\frac{1}{2}}\left\{(R'^2+L'^2n^2)^{\frac{1}{2}}(K^2+S^2n^2)^{\frac{1}{2}}\pm(KR'-L'Sn^2)\right\}^{\frac{1}{2}};$$

G_0, H_0, are terminal functions depending upon the apparatus at $z=0$; G_1, H_1, upon that at $z=l$; the apparatus being of any kind, specified by resistance-operators, making R_0', L_0' the effective resistance and inductance of apparatus at $z=0$, and R_1', L_1', at $z=l$. G_0 is given by

$$G_0=1+(R'^2+L'^2n^2)^{-1}[(P^2+Q^2)(R_0'^2+L_0'^2n^2)$$
$$+2P(R'R_0'+L'L_0'n^2)+2Qn(R_0'L'-R'L_0'),$$

from which H_0 is derived by changing the signs of P and Q; whilst G_1 and H_1 are the same functions of R_1', L_1' as G_0 and H_0 are of R_0', L_0'.

Now drop the accents, since we have only copper wires of low resistance (but not very thick) in question, and the terminal apparatus are to be of the simplest character. K/Sn will be vanishingly small practically, so take $K=0$. Next let R/Ln be small, and let the apparatus at $z=l$ be a mere coil, R_1, of negligible inductance first. We shall now have

$$P=R/2Lv, \qquad Q=n/v,$$

and these make

$$G_1^{\frac{1}{2}}=\left(1+\frac{R_1}{Lv}\right), \qquad H_1^{\frac{1}{2}}=\left(1-\frac{R_1}{Lv}\right).$$

Thus $R_1=Lv$ makes H_1 vanish, whatever the length of line, and the terms due to reflexions disappear.

We now have

$$C_0=\frac{V_0}{Lv}\epsilon^{-Rl/2Lv}\times G_0^{-\frac{1}{2}},$$

where $G_0^{-\frac{1}{2}}$ expresses the effect of the apparatus at $z=0$ in reducing the potential-difference there, V_0 being the impressed force, and the value of G_0 being unity where there is a short-circuit.

Now to show that $R_1=Lv$ makes the magnetic force of the receiver the greatest, go back to the general formula, let ϵ^{-Pl} be small, and let the size of wire vary, whilst the size of the receiving-coil is fixed. It will be easily found, from the expression for G_1, that the magnetic force of the coil is a maximum when

$$R_1^2+L_1^2n^2=\left(\frac{R^2+L^2n^2}{K^2+S^2n^2}\right)^{\frac{1}{2}},$$

where we keep in L_1 the inductance of the receiver. Or, when R/Ln and K/Sn are both small,

$$(R_1^2+L_1^2n^2)^{\frac{1}{2}}=Lv,$$

or, as described, $R_1=Lv$ when the receiver has a sufficiently small time-constant. The rule is, equality of impedances.

We may operate in a similar manner upon the terminal

function at the sending end. Suppose the apparatus to be representable as a resistance containing an electromotive force, and that by varying the resistance we cause the electromotive force to vary as its square root. Then, according to a well-known law, the arrangement producing the maximum external current is given by $R_0=Lv$, equality of impedances again. This brings us to

$$C_0=\frac{V_0}{2Lv}\epsilon^{-Rl/2Lv};$$

as if the circuit were infinitely long both ways, with maximum efficiency secured at both ends.

Lastly, the choice of L such that $Rl=2Lv$ makes the circuit, of given resistance, most efficient.

In long-distance telephony using wires of low resistance, the waves are sent along the circuit in a manner closely resembling the transmission of waves along a stretched elastic cord, subject to a small amount of friction. In order to similarly imitate the electrostatic theory, we must so reduce the mass of the cord, or else so exaggerate the friction, that there cannot be free vibrations. We may suppose that the displacement of the cord represents the potential-difference in both cases. But the current will be in the same phase as the potential-difference in one case, and proportional to its variation along the circuit in the other.

We may conveniently divide circuits, so far as their signalling peculiarities are concerned, into five classes. (1) Circuits of so short length, or so operated upon, that any effects due to electric displacement are insensible. The theory is then entirely electromagnetic, at least so far as numerical results are concerned. (2) Circuits of such great length that they can only be worked so slowly as to render electromagnetic inertia numerically insignificant in its effects. Also some telephonic circuits in which R/Ln is large. Then, at least so far as the reception of signals is concerned, we may apply the electrostatic theory. (3) The exceedingly large intermediate class in which both the electrostatic and electromagnetic sides have to be considered, not separately, but conjointly. (4) The simplified form of the last to which we are led when the signals are very rapid and the wires of low resistance. (5) The non-distortional circuit, in which, by a proper amount of uniform leakage, distortion of signals is abolished, whether fast or slow. Regarded from the point of view of practical application, this class lies on one side. But from the theoretical point of view, the non-distortional circuit lies in the very focus of the general theory, reducing it to simple algebra. I was led to it by an examination of the effect of telephones bridged across a common circuit (the proper place for intermediate apparatus, removing their impedance) on waves transmitted along the circuit. The current is reflected positively, the charge negatively, at a bridge. This is the opposite of what occurs when a resistance is put in the main circuit, which causes positive reflexion of the charge, and negative of the current. Unite the two effects and the reflexion of the wave is destroyed, approximately when the resistance in the main circuit and the bridge resistance are finite, perfectly when they are infinitely small, as in a uniform non-distortional circuit.

From the PHILOSOPHICAL MAGAZINE for May 1888.

On Electromagnetic Waves, especially in relation to the Vorticity of the Impressed Forces; and the Forced Vibrations of Electromagnetic Systems. By OLIVER HEAVISIDE.

[Continued from p. 156.]

Spherical Electromagnetic Waves.

15. LEAVING the subject of plane waves, those next in order of simplicity are the spherical. Here, at the very beginning, the question presents itself whether there can be anything resembling condensational waves?

Sir W. Thomson (Baltimore Lectures, as reported by Forbes in 'Nature," 1884) suggested that a conductor charged rapidly alternately + and − would cause condensational waves in the æther. But there is no other way of charging it than by a current from somewhere else, so he suggested two conducting spheres to be connected with the poles of an alternating dynamo. The idea seems to be here that electricity would be forced out of one sphere and into the other to and fro with great rapidity, and that between the spheres there might be condensational waves.

But in this case, according to the Faraday law of induction, the result would be the setting up of alternating electromagnetic disturbances in the dielectric, exposing the bounding surfaces of the two spheres to rapidly alternating magnetizing and electrizing force, causing waves, approximately spherical at least, to be transmitted into the spheres, in the diffusion manner, greatly attenuating as they progressed inward.

Perhaps, however, there can be condensational waves if we admit that a certain quite hypothetical something called electricity is compressible, instead of being incompressible, as it must be if we in Maxwell's scheme make the unnecessary assumption that an electric current is the motion through space of the something. In fact, Prof. J. J. Thomson has calculated* the speed of condensational waves supposed to arise by allowing the electric current to have convergence. But a careful examination of his equations will show that the condensational waves there investigated do not exist, *i. e.* the function determining them has the value zero†.

16. To construct a perfectly general spherical wave we may proceed thus. The characteristic equation of **H**, the magnetic force, in a homogeneous medium free from impressed force is, by (2) and (3),

$$\nabla^2\mathbf{H}=(4\pi\mu kp+\mu cp^2)\mathbf{H}. \quad . \quad . \quad . \quad . \quad (93)$$

Now, let **r** be the vector distance from the origin, and Q any *scalar* function satisfying this equation. Let

$$\mathbf{H}=\operatorname{curl}(\mathbf{r}Q). \quad . \quad . \quad . \quad . \quad . \quad . \quad (94)$$

Then this derived vector will satisfy (93), and have no convergence, and have no radial component, or will be arranged in spherical sheets. From it derive the other electromagnetic quantities. Change **H** to **E** to obtain spherical sheets of electric force.

This method leads to the spherical sheets depending upon any kind of spherical harmonic. They are, however, too general to be really useful except as mathematical exercises. For the examination of the manner of origin and propagation

* B. A. Report on Electrical Theories.

† I ought to qualify this by adding that the investigation seems very obscure, so that although I cannot make the system work, yet others may

of waves, zonal harmonics are more useful, besides leading to the solution of more practical problems. It is then not difficult to generalize results to suit any kind of spherical harmonic.

17. *The simplest Spherical Waves.*—Let the lines of **H** be circles, centred upon the axis from which θ is measured, and let r be the distance from the origin. We have no concern with ϕ (longitude) as regards **H**, so that the simple specification of its intensity H fully defines it. Under these circumstances the equation (93) becomes

$$\left.\begin{aligned}(r\mathrm{H})'' + \frac{\nu}{r^2}(\nu r\mathrm{H})^{``} &= (4\pi\mu_0 kp + \mu_0 cp^2)\,\mathrm{H}, \\ &= q^2\mathrm{H}, \text{ say,}\end{aligned}\right\} \quad . \quad (95)$$

where the acute accent denotes differentiation to r, and the grave accent to $\cos\theta$ or μ, whilst ν stands for $\sin\theta$. The inductivity will be now μ_0, to avoid confusing with the μ of zonal harmonics. Equation (95) also defines q in the three forms it can assume in a conductor, dielectric, and conducting dielectric.

Now try to make of r**H** an undistorted spherical wave, *i. e.* H varying inversely as the distance, and travelling inward or outward at speed v. Let

$$r\mathrm{H} = \mathrm{A}f(r - vt), \quad . \quad . \quad . \quad . \quad . \quad (96)$$

where A is independent of r and t. Of course we must have $k=0$, making $q=p/v$. Now (96) makes

$$v^2(r\mathrm{H})'' = rp^2\mathrm{H}\,; \quad . \quad . \quad . \quad . \quad . \quad (97)$$

which, substituted in (95), gives

$$\nu(\nu\mathrm{H})^{``} = 0\,; \quad . \quad . \quad . \quad . \quad . \quad . \quad . \quad . \quad (98)$$

therefore

$$\mathrm{A}\nu = \mathrm{A}_1\mu + \mathrm{B}_1. \quad . \quad . \quad . \quad . \quad . \quad . \quad . \quad (99)$$

From these we find the required solutions to be

$$\mathrm{H} = \mathrm{E}/\mu_0 v = \frac{\mathrm{A}_1\mu + \mathrm{B}_1}{r\nu}\mathrm{F}_0'(r - vt), \quad . \quad . \quad . \quad (100)$$

$$\mathrm{F} = \mu_0 v\,\frac{\mathrm{A}_1}{r^2}\mathrm{F}_0(r - vt)\,; \quad . \quad . \quad . \quad . \quad . \quad . \quad . \quad (101)$$

where F_0 is any function, A_1 and B_1 constants, E and F the two components of the electric force, F being the radial component out, and E the other component coinciding with a line of longitude, the positive direction being that of increasing θ, or from the pole. Similarly, if the lines of **E** be circular about the axis, we have the solutions

$$E = -\mu_0 v H_\theta = -\mu_0 v \frac{A_1\mu + B_1}{r\nu} F_0'(r-vt), \quad . \quad (102)$$

$$H_r = \frac{A_1}{r^2} F_0(r-vt), \quad . \quad . \quad . \quad . \quad . \quad (103)$$

where H_r and H_θ are the radial and tangential components of **H**.

But both these systems involve infinite values at the axis. We must therefore exclude the axis somehow to make use of them. Here is one way. Describe a conical surface of any angle θ_1, and outside it another of angle θ_2, and let the dielectric lie between them. Make the tangential component of **E** at the conical surfaces vanish, requiring infinite conductivity there, and we make F vanish in (101), and produce the solution

$$E = \mu_0 v H = \frac{B}{r\nu} f(r-vt), \quad . \quad . \quad . \quad . \quad (104)$$

exactly resembling plane waves as regards $r\nu E$. Here B is the same as $\mu_0 v B_1$, and f the same as F_0', in equation (100) *.

18. Now bring in zonal harmonics. Split equation (95) into the two

* In order to render this arrangement (104) intelligible in terms of more everyday quantities, let the angles θ_1 and θ_2 be small, for simplicity of representation; then we have two infinitely conducting tubes of gradually increasing diameter enclosing between them a non-conducting dielectric. Now change the variables. Let V be the line-integral of **E** across the dielectric, following the direction of the force; it is the potential-difference of the conductors. Let $4\pi C$ be the line-integral of H round the inner tube; it is the same for a given value of r, independent of θ; C is therefore what is commonly called the current in the conductor. We shall have

$$V = LvC, \quad C = SvV, \quad LSv^2 = 1;$$

where L is the inductance and S the permittance, per unit length of the circuit. The value of L is

$$L = 2\mu_0 \log\left[(\tan\tfrac{1}{2}\theta_2) \div (\tan\tfrac{1}{2}\theta_1)\right];$$

so that the circuit has uniform inductance and permittance. The value of C in terms of (104) is

$$C = \frac{B}{2\mu_0 v} f(r-vt).$$

When the tubes have constant radii a_1 and a_2, the value of L reduces to the well known

$$L = 2\mu_0 \log(a_2/a_1),$$

of concentric cylinders. The wave may go either way though only the positive wave is mentioned.

$$(r\mathrm{H})'' = \left\{ q^2 + \frac{m(m+1)}{r^2} \right\} r\mathrm{H}, \quad . \quad . \quad . \quad (105)$$

$$\frac{\nu}{r^2}(\nu\mathrm{H})^{\prime\prime} = -\frac{m(m+1)}{r^2}\mathrm{H}. \quad . \quad . \quad . \quad . \quad (106)$$

The equation (106) has for solution

$$\mathrm{H} = \mathrm{A}\nu\mathrm{Q}'_m,$$

where A is independent of θ, and is to be found from (105).

The most practical way of getting the r functions is that followed by Professor Rowland in his paper*, wherein he treats of the waves emitted when the state is sinusoidal with respect to the time. We shall come across the same waves in some problems.

Let

$$\mathrm{H} = \mathrm{P}_m \frac{\epsilon^{qr}}{r} \nu\mathrm{Q}'_m. \quad . \quad . \quad . \quad . \quad . \quad . \quad (107)$$

Then the equation of P_m is, by insertion of (107) in (105),

$$\mathrm{P}'' + 2q\mathrm{P}' = \frac{m(m+1)}{r^2}\mathrm{P}; \quad . \quad . \quad . \quad . \quad (108)$$

and the solution, for practical purposes with complete harmonics, is

$$\mathrm{P} = 1 - \frac{m(m+1)}{2qr} + \frac{m(m^2-1^2)(m+2)}{2\,.\,4q^2r^2} - \frac{m(m^2-1)(m^2-2^2)(m+3)}{2.4.6q^3r^3} + \dots \quad . \quad (109)$$

We shall find the first few useful, thus :—

$$\left.\begin{aligned} \mathrm{P}_1 &= 1 - (qr)^{-1}, \\ \mathrm{P}_2 &= 1 - 3(qr)^{-1} + 3(qr)^{-2}, \\ \mathrm{P}_3 &= 1 - 6(qr)^{-1} + 15(qr)^{-2} - 15(qr)^{-3}. \end{aligned}\right\} \quad . \quad (110)$$

Now let $\mathrm{U} = \epsilon^{qr}\mathrm{P}$, so that U is the r function in $\mathrm{H}r$. If we change the sign of q in U, producing, say, W, it is the required second solution of (105). Thus

$$\mathrm{U}_1 = \epsilon^{qr}\left(1 - \frac{1}{qr}\right), \quad \mathrm{W}_1 = \epsilon^{-qr}\left(1 + \frac{1}{qr}\right), \quad . \quad . \quad . \quad (111)$$

in the very important case of Q_1, when $m = 1$.

* Phil. Mag. June 1884, "On the Propagation of an Arbitrary Electromagnetic Disturbance, Spherical Waves of Light, and the Dynamical Theory of Refraction." Prof. J. J. Thomson has also considered spherical waves in a dielectric in his paper "On Electrical Oscillations and Effects produced by the Motion of an Electrified Sphere," Proc. London Math. Soc. vol. xv., April 3, 1884.

The conjugate property of U and W is

$$UW'-U'W=-2q, \quad \ldots \ldots \quad (112)$$

which is continually useful.

We have next to combine U and W so as to produce functions suitable for use inside spheres, right up to the centre, and finite there. Let

$$u=\tfrac{1}{2}(U+W), \quad w=\tfrac{1}{2}(U-W). \quad \ldots \quad (113)$$

It will be found that when m is even, w/r is zero and u/r infinite at the origin; but that when m is odd, it is u/r that is zero at the origin and w infinite.

The conjugate property of u and w is

$$uw'-u'w=q, \quad \ldots \ldots \quad (114)$$

corresponding to (112).

19. *Construction of the Differential Equations connected with a Spherical Sheet of Vorticity of Impressed Force.*—Now let there be two media—one extending from $r=0$ to $r=a$, in which we must therefore use the u function or w function, according as m is odd or even, and an outer medium, or at least one in which q has a different form in general. Then, within the sphere of radius a, we have

$$H=Ar^{-1}u, \quad \ldots \ldots \quad (115)$$

$$-k_1E=Ar^{-1}u', \quad \ldots \ldots \quad (116)$$

where $k_1=4\pi k+cp$, and we suppose m odd. It follows that

$$\frac{E}{H}=-\frac{1}{k_1}\frac{u'}{u}. \quad \ldots \ldots \quad (117)$$

In the outer medium use W, if the medium extends to infinity, or both U and W if there be barriers or change of medium. First, let it be an infinitely extended medium. Then, in it,

$$H=Br^{-1}(u-w), \quad \ldots \ldots \quad (118)$$

$$-k_2E=Br^{-1}(u'-w'), \quad \ldots \ldots \quad (119)$$

where $k_2=4\pi k+cp$ in the outer medium. From these

$$\frac{E}{H}=-\frac{1}{k_2}\frac{u'-w'}{u-w}. \quad \ldots \ldots \quad (120)$$

(117) and (120) show the forms of the resistance-operators on the two sides *.

* Some rather important considerations are presented here. On what principles should we settle which functions to use, internally and externally, seeing that these functions U and W are not quantities, but differential operators? First, as regards the space outside the surface of origin

Now, at the surface of separation, $r=a$, H is continuous (unless we choose to make it a sheet of electric current, which we do not); so that the H in (117) and in (120) are the same. We only require a relation between the E's to complete the differential equation.

Let there be vorticity of impressed force on the surface $r=a$, and nowhere else (the latter being already assumed). Then

$$\text{curl}\,\mathbf{e} = \text{curl}\,\mathbf{E} \quad . \quad . \quad . \quad . \quad . \quad . \quad (121)$$

is the surface-condition which follows; or, if f be the measure of the curl of **e**,

$$f = E_2 - E_1, \quad . \quad . \quad . \quad . \quad . \quad . \quad (122)$$

E_2 meaning the outer and E_1 the inner E. Therefore

$$f = H_a\left(\frac{E_2}{H_2} - \frac{E_1}{H_1}\right), \quad . \quad . \quad . \quad . \quad . \quad (123)$$

H_a denoting the surface H. So, by (117) and (120), used in (123),

$$f = \left(\frac{1}{k_1}\frac{u_1'}{u_1} - \frac{1}{k_2}\frac{u_2'-w_2'}{u_2-w_2}\right) H_a, \quad (r=a), \quad . \quad . \quad (124)$$

the required differential equation. Observe that u_1 only differs from u_2 and w_1 from w_2 in the different values of q inside and outside (when different), and that $r=a$ in all.

---

of disturbances. The operator ϵ^{qr} turns $f(t)$ into $f(t+r/v)$, and can therefore only be possible with a negative wave, coming to the origin. But there cannot be such a wave without a barrier or change of medium to produce it. Hence the operator ϵ^{-qr} alone can be involved in the external solution when the medium is unbounded, and we must use W. Next, go inside the sphere $r=a$. It is clear that both U and W are now needed, because disturbances come to any point from the further as well as from the nearer side of the surface, thus coming from and going to the centre. Two questions remain: Why take U and W in equal ratio? ; and why their sum or their difference, according as m is odd or even? The first is answered by stating the facts that, although it is convenient to assume the origin to be a place of reflexion, yet it is really only a place where disturbances cross, and that the H produced at any point of the surface is (initially) equal on both sides of it. The second question is answered by stating the property of the Q'_m function, that it is an even function of μ when m is odd, and conversely; so that when m is odd the H disturbances arriving at any point on a diameter from its two ends are of the same sign, requiring U+W ; and when m is even, of opposite signs, requiring U−W.

Similar reasoning applies to the operators concerned in other than spherical waves. Cases of simple diffusion are brought under the same rules by generalizing the problem so as to produce wave-propagation with finite speed. On the other hand, when there are barriers, or changes of media, there is no difficulty, because the boundary conditions tell us in what ratio U and W must be taken.

Equation (124) applies to any odd m. When m is even, exchange u and w, also u' and w'. In the mth system we may write

$$f_m = \phi_m \mathrm{H}_a, \quad . \quad . \quad . \quad . \quad . \quad . \quad (125)$$

the form of ϕ being given in (124). The vorticity of the impressed force is of course restricted to be of the proper kind to suit the mth zonal harmonic. Thus, any distribution of vorticity whose lines are the lines of latitude on the spherical surface may be expanded in the form

$$\Sigma f_m \nu \mathrm{Q}_m', \quad . \quad . \quad . \quad . \quad . \quad . \quad . \quad (126)$$

and it is the mth of these distributions which is involved in the preceding.

20. Both media being supposed to be identical, ϕ reduces to

$$\phi = \frac{1}{k_1} \frac{q}{u_a(u_a - w_a)}, \quad . \quad . \quad . \quad . \quad . \quad . \quad (127)$$

by using (114) in (124). This is with m odd; if even, we shall get

$$\phi = \frac{1}{k_1} \frac{q}{w_a(u_a - w_a)}. \quad . \quad . \quad . \quad . \quad . \quad . \quad (128)$$

In a non-dielectric conductor, $k_1 = 4\pi k$, and $q^2 = 4\pi\mu k p$; so that, keeping to m odd,

$$\phi = \left(\frac{\mu p}{4\pi k}\right)^{\frac{1}{2}} \frac{1}{u_a(u_a - w_a)}. \quad . \quad . \quad . \quad . \quad . \quad (129)$$

In a non-conducting dielectric, $k_1 = cp$, and $q = p/v$; so

$$\phi = \frac{\mu_0 v}{u_a(u_a - w_a)}. \quad . \quad . \quad . \quad . \quad . \quad . \quad . \quad (130)$$

In this case the complete differential equation is

$$\mathrm{H}_a = \Sigma \frac{\nu \mathrm{Q}_m'}{\mu_0 v} u_a(u_a - w_a) f_m, \quad . \quad . \quad . \quad . \quad (131)$$

when there is any distribution of impressed force in space whose vorticity is represented by (126).

Outside the sphere, consequently

$$(\text{out}) \left\{ \begin{aligned} \mathrm{H} &= \Sigma \frac{\nu \mathrm{Q}_m'}{\mu_0 v} \frac{a}{r} u_a(u - w) f_m, \quad . \quad . \quad . \quad (132) \\ -cp\mathrm{E} &= \Sigma \frac{\nu \mathrm{Q}_m'}{\mu_0 v} \frac{a}{r} u_a(u' - w') f_m, \quad . \quad . \quad . \quad (133) \end{aligned} \right.$$

understanding that when no letter is affixed to u or w, the value at distance r is meant. We see at once that $u_a = 0$ makes the external field vanish, *i. e.* the field of the particular

f concerned. This happens, when f is a sinusoidal function of the time, at definite frequencies. Also, inside the sphere,

$$\text{(in)} \quad \begin{cases} H = \Sigma \dfrac{\nu Q'_m}{\mu_0 v} \dfrac{a}{r} u(u_a - w_a) f_m, & \quad (134) \\ -cpE = \Sigma \dfrac{\nu Q'_m}{\mu_0 v} \dfrac{a}{r} u'(u_a - w_a) f_m. & \quad (135) \end{cases}$$

As for the radial component F, it is not often wanted. It is got thus from H:—

$$-cpF = \frac{1}{r}(\nu H)', \quad . \quad . \quad . \quad . \quad . \quad . \quad (136)$$

where for cp write $4\pi k + cp$ in the general case. Thus, the internal F corresponding to (135) is

$$\text{(in)} \qquad cpF = \Sigma \frac{m(m+1)}{\mu_0 v} \frac{a}{r^2} u(u_a - w_a) f_m Q_m. \quad (137)$$

21. *Practical Problem. Uniform Impressed Force in the Sphere.*—If there be a uniform field of impressed force in the sphere, parallel to the axis, of intensity f_1, its vorticity is represented by $f_1 \sin\theta$ on the surface of the sphere. It is therefore the case $m=1$ in the above. Let this impressed force be suddenly started. Find the effect produced. We have, by (132) *,

$$\text{(out)} \qquad H = u_a(u-w)\frac{f_1 \nu a}{\mu_0 v r}; \quad . \quad . \quad . \quad . \quad . \quad (138)$$

* It will be observed that the operator connecting f_1 and H is of such a nature that the process of expansion of H in a series of normal functions fails. I have examined several cases of this kind. The invariable rule seems to be that when there is a surface of vorticity of **e**, leading to an equation of the form $f=\phi H$, and there is a change of medium somewhere, or else barriers, causing reflected waves, the form of ϕ is such that we can, when f is constant, starting at $t=0$, solve thus

$$H = \frac{f}{\phi_0} + \Sigma \frac{f\epsilon^{pt}}{p(d\phi/dp)},$$

extending over all the (algebraical) p roots of $\phi=0$, which is the determinantal equation. But should there be no change of medium, the conjugate property of the functions concerned comes into play. It causes a great simplification in the form of ϕ, and makes the last method fail completely, all traces of the roots having disappeared. But if we pass continuously from one case to the other, then the last formula becomes a definite integral. On the other hand, we can immediately integrate $f=\phi H$ in its simplified form, and obtain an interpretable equivalent for the definite integral, which latter is more ornamental than useful. In the simplified form, ϕ may be either rational or irrational. The integration of the irrational forms will be given in some later problems.

or, in full, referring to the forms of u and w, equations (110) to (113),

$$\mathrm{H}=\frac{va}{2\mu_0 vr}\left\{\epsilon^{-q(r-a)}\left(1-\frac{1}{qa}\right)\left(1+\frac{1}{qr}\right)\right.$$

$$\left.+\epsilon^{-q(r+a)}\left(1+\frac{1}{qa}\right)\left(1+\frac{1}{qr}\right)\right\}f_1. \quad . \quad (139)$$

Effect the integrations indicated by the inverse powers of q or p/v; thus

$$\frac{f_1}{q^n}=f_1\frac{(vt)^n}{\lfloor n}, \quad . \quad . \quad . \quad . \quad . \quad . \quad . \quad (140)$$

if f_1 be zero before and constant after $t=0$. As for the exponentials, use Taylor's theorem, as only differentiations are involved. We get, after the process (140) has been applied to (139), and then Taylor's theorem carried out,

$$\mathrm{H}=\frac{f_1 va}{2\mu_0 vr}\left\{\left(1-\frac{vt_1}{a}+\frac{vt_1}{r}-\frac{v^2t_1{}^2}{2ar}\right)+\left(1+\frac{vt_2}{a}+\frac{vt_2}{r}+\frac{v^2t_2{}^2}{2ar}\right)\right\}, \quad (141)$$

where

$$vt_1=vt-r+a,$$

$$vt_2=vt-r-a.$$

It is particularly to be noticed that the t_1 part of (141) only comes into operation when t_1 reaches zero, and similarly as regards the t_2 part. Thus, the first part expresses the primary wave out from the surface; the second, arriving at any point $2a/v$ later than the first, is the reflected wave from the centre, arising from the primary wave inward from the surface.

The primary wave outward may be written

$$\mathrm{H}=\left(\frac{f_1\nu}{2\mu_0 v}\right)\frac{1}{2}\left(1+\frac{a^2-v^2t^2}{r^2}\right), \quad . \quad . \quad . \quad (142)$$

where $vt>(r-a)$, and the second wave by its exact negative, with $vt>(r+a)$. Now, by comparing (132) with (134), we see that the internal solution is got from the external by exchanging a and r in the $\{\}$'s in (139) and (141), including also in t_1 and t_2. The result is that (142) represents the internal H in the primary inward wave, vt having to be $>(a-r)$; whilst its negative represents the reflected wave, provided $vt>(a+r)$.

The whole may be summed up thus. First, vt is $<a$. Then (142) represents H everywhere between $r=a+vt$ and $r=a-vt$. But when vt is $>a$, H is given by the same formula between the limits $r=vt-a$ and $vt+a$. In both cases H is zero outside the limits named.

The reflected wave, superimposed on the primary, annuls the H disturbance, which is therefore, after the reflexion, confined to a spherical shell of depth $2a$ containing the uncancelled part of the primary wave outward.

The amplitude of H at the fronts of the two primary waves, in and out, before the former reaches the centre, is

$$(f_1 \nu a) \div (2\mu_0 vr).$$

After the inward wave has reached the centre, however, the amplitude of H on the front of the reflected wave is the negative of that of the primary wave at the same distance, which is itself negative.

The process of reflexion is a very remarkable one, and difficult to fully understand. At the moment $t = a/v$ that the disturbance reaches the centre, we have $\mathrm{H} = (f_1 \nu) \div (4\mu_0 v)$, constant, all the way from $r=0$ to $2a$, which is just half the initial value of H on leaving the surface of the sphere. But just before reaching the centre, H runs up infinitely for an infinitely short time, infinitely near the centre; and just after the centre is reached we have $\mathrm{H} = -\infty$ infinitely near the centre, where the H disturbance is always zero, except in this singular case when it is seemingly finite for an infinitely short time, though, of course, ν is indeterminate.

With respect to this running-up of the value of H in the inward primary wave, it is to be observed that whilst H is increasing so fast at and near its front, it is falling elsewhere, viz. between near the front and the surface of the sphere; so that just before the centre is reached H has only half the initial value, except close to the centre, where it is enormously great.

After reflexion has commenced, the H disturbance is negative in the hinder part of the shell of depth $2a$ which goes out to infinity, positive of course still in the forward part. At a great distance these portions become of equal depth a; at its front $\mathrm{H} = (f_1 \nu a)(2\mu_0 vr)^{-1}$, at its back $\mathrm{H} = -$ ditto; using of course a different value of r.

22. As regards the electric field, we have, by (133),

$$\text{(out)} \qquad \mathrm{E} = -\frac{1}{cp}\frac{\nu}{\mu_0 v}\frac{a}{r} u_a(u' - w')f_1 ; \quad . \quad . \quad (143)$$

which, expanded, is

$$\mathrm{E} = \frac{\nu a}{2r}\left\{ \epsilon^{q(a-r)}\left(1 - \frac{1}{qa}\right)\left(1 + \frac{1}{qr} + \frac{1}{q^2 r^2}\right) \right.$$

$$\left. + \epsilon^{-q(a+r)}\left(1 + \frac{1}{qa}\right)\left(1 + \frac{1}{qr} + \frac{1}{q^2 r^2}\right) \right\} f_1 ; \quad . \quad (144)$$

comparing which with (139), we see that

$$E=\mu vH+\frac{va}{2r}\left\{\epsilon^{-q(r-a)}\frac{1}{q^2r^2}\left(1-\frac{1}{qa}\right)+\epsilon^{-q(r+a)}\frac{1}{q^2r^2}\left(1+\frac{1}{qa}\right)\right\}f_1. \quad (145)$$

We have, therefore, only to develop the second part, which is not in the same phase with H. It is, in the same manner as before,

$$\frac{f_1va}{2r}\left(\frac{v^2t_1^{\,2}}{2r^2}-\frac{v^3t_1^{\,3}}{6r^2a}\right)+\frac{f_1va}{2r}\left(\frac{v^2t_2^{\,2}}{2r^2}+\frac{v^3t_2^{\,3}}{6r^2a}\right), \quad . \quad . \quad . \quad (146)$$

only operating when $vt_1=vt-r+a$, and $vt_2=vt-r-a$ are positive. Or,

$$\frac{f_1va}{4r^3}\left\{\left(\frac{2a^2}{3}+a(vt-r)-\frac{(vt-r)^3}{3a}\right)_1 + \left(\frac{2a^2}{3}-a(vt-r)+\frac{(vt-r)^3}{3a}\right)_2\right\}, \quad (146\,a)$$

1 and 2 referring to the two waves. So, when $vt>(r+a)$, and the two are coincident, we have the sum

$$E=\frac{f_1va^3}{3r^3}, \quad . \quad . \quad . \quad . \quad . \quad . \quad . \quad (147)$$

which is the tangential component of the steady electric field left behind.

The radial component F is, by (137),

$$\text{(out)} \quad F=\frac{\cos\theta}{r}\left\{\epsilon^{-q(r-a)}\left(\frac{1}{qr}+\frac{1}{q^2r^2}-\frac{1}{q^2ra}-\frac{1}{q^3r^2a}\right)+\ldots\right\}f_1, \quad (148)$$

where the unwritten term ... may be obtained from the preceding by changing the sign of a. Or,

$$F=\frac{f_1a\cos\theta}{r^2}\left\{\left(vt_1+\frac{v^2t_1^{\,2}}{2r}-\frac{v^2t_1^{\,2}}{2a}-\frac{v^3t_1^{\,3}}{6ra}\right)+\ldots\right\}, \quad (149)$$

where $vt_1=vt+a-r$. Or,

$$F=\frac{f_1a\cos\theta}{r^2}\left\{\frac{a^2}{3r}+\frac{a}{2}+\frac{a}{2r}(vt-r)-\frac{1}{2a}(vt-r)^2 - \frac{1}{6ra}(vt-r)^3+\ldots\right\}; \quad (150)$$

so that, when both waves coincide, we have their sum,

$$F=\frac{2f_1a^3\cos\theta}{3r^3}, \quad . \quad . \quad . \quad . \quad . \quad . \quad (151)$$

which is the radial component of the steady field left behind by the part of the primary wave whose magnetic field is wholly cancelled.

To verify, the uniform field of impressed force of intensity f_1, by elementary principles, produces the external electric potential

$$\Omega = f_1 \cos\theta \frac{a^3}{3r^2},$$

whose derivatives, radial and tangential, taken negatively, are (151) and (147). The corresponding internal potential is

$$\Omega = \tfrac{1}{3} f_1 r \cos\theta.$$

But its slope does not give the force **E** left behind within the sphere, because this **E** is the force of the flux. Any other distribution of impressed force, with the same vorticity, will lead to the same **E**. Our equation (135) and its companion for F, derived from (134) by using (136), lead to the steady field (residual)

$$\mathrm{E} = -\tfrac{2}{3} f_1 \sin\theta, \quad \mathrm{F} = \tfrac{2}{3} f_1 \cos\theta, \quad . \quad . \quad . \quad (152)$$

the components of the true force of the flux. Add **e** to the slope of Ω to produce **E** *.

F is always zero at the front of the primary wave outward, and $\mathrm{E} = \mu_0 v \mathrm{H}$. At the front of the primary wave inward F is also zero, and $\mathrm{E} = -\mu_0 v \mathrm{H}$. After reflexion, F at the front of the reflected wave is still zero, but now $\mathrm{E} = \mu_0 v \mathrm{H}$.

The electric energy U_1 set up is the volume-integral of the scalar product $\frac{1}{2}$**eD**. That is,

$$\mathrm{U}_1 = \tfrac{1}{2} f_1 \times \tfrac{2}{3} \frac{c f_1}{4\pi} \times \frac{4\pi a^3}{3} = \frac{c a^3 f_1^2}{9}. \quad . \quad . \quad . \quad (153)$$

But the total work done by **e** is $2\mathrm{U}_1$, by the general law that the whole work done by impressed forces suddenly started exceeds the amount representing the waste by Joule heating at the final rate (when there is any), supposed to start at once, by twice the excess of the electric over the magnetic energy of the steady field set up. It is clear, then, that when the travelling shell has gone a good way out, and it has become nearly equivalent to a plane wave, its electric and magnetic energies are nearly equal, and each nearly $\frac{1}{2}\mathrm{U}_1$

* Sometimes the flux is apparently wrongly directed. For example, a uniform field of impressed force from left to right in all space except a spherical portion produces a flux from right to left in that portion. This is made intelligible by the above. Let the impressed force act in the space between $r=a$ and $r=b$, a being small and b great. In the inner sphere the first effects are those due to the $r=a$ vorticity, and the flux left behind is against the force. But after a time comes the wave from the $r=b$ vorticity, which sets matters right. The same applies in the case of conductors, when, in fact, a long time might have to elapse before the second and real permanent state conquered the first one.

in value. I did not, however, anticipate that the magnetic energy in the travelling shell would turn out to be constant, viz. $\frac{1}{2}U_1$ during the whole journey, from $t=a/v$ to $t=\infty$, so that it is the electric energy in the shell which gradually decreases to $\frac{1}{2}U_1$. Integrate the square of H according to (142) to verify.

23. The most convenient way of reckoning the work done, and also the most appropriate in this class of problems, is by the integral of the scalar product of the curl of the impressed force and the magnetic force. Thus, in our problem

$$2U_1 = \int dt\Sigma \mathbf{e}\Gamma = \int dt\Sigma \mathbf{H}\ \text{curl}\ \mathbf{e}/4\pi$$

$$= \iint\frac{f_1\nu dS}{4\pi}\int H_a dt, \quad . \quad . \quad . \quad . \quad . \quad (154)$$

where dS is an element of the surface $r=a$. So we have to calculate the time-integral of the magnetic force at the place of vorticity of $\mathbf{e}$, the limits being 0 and $2a/v$. This can be easily done without solving the full problem, not only in the case of $m=1$, but $m=$ any integer. The result is, if U_m be the electric energy of the steady field due to f_m,

$$\int H_a dt = \frac{caf_m\nu\,\dot{Q}_m}{2m+1}, \quad . \quad . \quad . \quad . \quad (155)$$

and, therefore, by surface-integration according to (154),

$$2U_m = a^3 cf_m^2\frac{m(m+1)}{2m+1}. \quad . \quad . \quad . \quad . \quad (156)$$

$\frac{1}{2}U_m$ is the magnetic energy in the mth travelling shell. I have entered into detail in the case of $m=1$, because of its relative importance, and to avoid repetition. In every case the magnetic field of the primary wave outward is cancelled by that of the reflexion of the primary wave inward, producing a travelling shell of depth $2a$, within which is the final steady field. There are, however, some differences in other respects, according as m is even or odd.

Thus, in the case $m=2$, we have, by (110) to (113),

$$\tfrac{1}{2}(U_a - W_a)W = \tfrac{1}{2}\Big\{\epsilon^{q(a-r)}\Big(1-\frac{3}{qa}+\frac{3}{q^2a^2}\Big)$$

$$-\epsilon^{-q(r+a)}\Big(1+\frac{3}{qa}+\frac{3}{q^2a^2}\Big)\Big\}\times\Big(1+\frac{3}{qr}+\frac{3}{q^2r^2}\Big). \quad (157)$$

Making this operate upon f_2, zero before and constant after $t=0$, we obtain, by (132), (140), and Taylor's theorem,

$$\text{(out)}\ H = \frac{f_2 a\nu \dot{Q}_2}{2\mu_0 vr}\Big\{\frac{1}{4}+\frac{3}{8}\Big(\frac{a^2}{r^2}+\frac{r^2}{a^2}\Big)-v^2t^2\frac{3}{4}\Big(\frac{1}{r^2}+\frac{1}{a^2}\Big)+\frac{3}{8}\frac{v^4t^4}{a^2r^2}-\ldots\Big\}\ (158$$

In the wave represented, $vt > (r-a)$, it being the primary wave out. The unrepresented part, to be obtained by changing the sign of a within the $\{\}$, is the reflected wave, in which $vt > (r+a)$.

To obtain the internal H exchange a and r within the $\{\}$ in (158). The result is that

$$\mathrm{H} = \frac{f_2 a v \mathrm{Q}'_2}{2\mu_0 v r}\left\{-\tfrac{1}{2} + \frac{3}{8a^2r^2}(v^2t^2 - a^2 - r^2)^2\right\} \quad . \quad (159)$$

expresses the H solution always, provided that when $vt < a$ the limits for r are $a - vt$ and $a + vt$; but when $vt > a$, they are $vt - a$ and $vt + a$.

At the surface of the sphere,

$$\mathrm{H}_a = \frac{f_2 \nu \mathrm{Q}'_2}{2\mu_0 v}\left\{1 - \frac{3}{2}\left(\frac{vt}{a}\right)^2 + \frac{3}{8}\left(\frac{vt}{a}\right)^4\right\}, \quad . \quad (160)$$

from $t=0$ to $2a/v$. It vanishes twice, instead of only once, intermediately, finishing at the *same* value that it commenced at, instead of at the opposite, as in the $m=1$ case.

The radial component F of **E** is always zero at the front of either of the primary waves or of the reflected wave, and $\mathrm{E} = \pm\mu_0 v\mathrm{H}$, according as the wave is going out or in. In the travelling shell H changes sign m times, thus making $m+1$ smaller shells of oppositely directed magnetic force. At its outer boundary

$$\mathrm{E} = \mu_0 v \mathrm{H} = \tfrac{1}{2} f_m \nu \mathrm{Q}'_m (a/r), \quad . \quad . \quad . \quad (161)$$

and at the inner boundary the same formula holds, with $\pm$ prefixed according as m is even or odd.

In case $m=3$, the magnetic force at the spherical surface is

$$\mathrm{H}_a = \frac{f_3 \nu \mathrm{Q}'_3}{2\mu_0 v}\left\{1 - \frac{3v^2t^2}{a^2} + \frac{15}{8}\frac{v^4t^4}{a^4} - \frac{5}{16}\frac{v^6t^6}{a^6}\right\} \quad . \quad (162)$$

from $t=0$ to $2a/v$, after which, zero.

24. *Spherical Sheet of Radial Impressed Force.*—If the surface $r=a$ be a sheet of radial impressed force, it is clear that the vorticity is wholly on the surface. Let the intensity be independent of ϕ, so that

$$e = \Sigma e_m \mathrm{Q}_m. \quad . \quad . \quad . \quad . \quad . \quad . \quad (163)$$

The steady potential produced is

$$\text{(in)} \qquad \mathrm{V}_1 = -\Sigma e_m \mathrm{Q}_m \frac{m+1}{2m+1}\left(\frac{r}{a}\right)^m. \quad . \quad . \quad (164)$$

$$\text{(out)} \qquad \mathrm{V}_2 = +\Sigma e_m \mathrm{Q}_m \frac{m}{2m+1}\left(\frac{a}{r}\right)^{m+1}, \quad (165)$$

because, at $r=a$, these make

$$V_2-V_1=e, \quad \text{and} \quad \frac{dV_1}{dr}=\frac{dV_2}{dr}; \quad . \quad . \quad (166)$$

i. e. potential-difference e and continuity of displacement. The normal component of displacement is

$$-\frac{c}{4\pi}\frac{dV_1}{dr}=\frac{c}{4\pi a}\Sigma e_m Q_m\frac{m(m+1)}{2m+1}; \quad . \quad . \quad .$$

therefore, integrating over the sphere, the total work done by e is

$$2U=\Sigma\, cae_m^2\frac{m(m+1)}{(2m+1)^2}, \quad . \quad . \quad . \quad . \quad (168)$$

which agrees with the estimate (156), because

$$f=-\frac{de}{ad\theta}=\frac{\nu}{a}\frac{de}{d\mu}, \quad . \quad . \quad . \quad . \quad (169)$$

finds the vorticity, f, from the radial impressed force e; or, taking

$$e=e_m Q_m$$

$$\therefore \quad \frac{e_m \nu Q'_m}{a}=\text{vorticity},$$

so that the old $f_m=e_m/a$.

25. *Single Circular Vortex Line.*—There are some advantages connected with transferring the impressed force to the surface of the sphere, as it makes the force of the flux and the force of the field identical both outside and inside. At the boundary F is continuous, E discontinuous.

Let the impressed force be a simple circular shell of radius a, and strength e. Let it be the equatorial plane, so that the equator is the one line of vorticity. Substitute for this shell a spherical shell of strength $\frac{1}{2}e$ on the positive hemisphere, $-\frac{1}{2}e$ on the negative, the impressed force acting radially. Expand this distribution in zonal harmonics. The result is

$$\Sigma e_m Q_m=\frac{e}{2}\left\{\frac{3}{2}Q_1-\frac{7}{2.4}Q_3+\frac{11.3}{2.4.6}Q_5\right.$$
$$\left.-\frac{15.1.3.5}{2.4.6.8}Q_7+\ldots\right\}, \quad . \quad . \quad (170)$$

so that we are only concerned with the odd m's. This equation settling the value of e_m, the vorticity is

$$\Sigma\frac{e_m}{a}\nu Q'_m=\Sigma f_m \nu Q'_m. \quad . \quad . \quad . \quad . \quad (171)$$

We know therefore by the preceding, the complete solution due to sudden starting of the single vortex line. That is, we know the individual waves in detail produced by e_1, e_3, &c. The resultant travelling disturbance is therefore confined between two spherical surfaces of radii $vt-a$ and $vt+a$, after the centre has been reached, or of radii $a-vt$ and $a+vt$ before the centre is reached. But it cannot occupy the whole of either of the regions mentioned.

The actual shape of the boundaries, however, may be easily found. It is sufficient to consider a plane section through the axis of the sphere. Let A and B be the points on this plane cut by the vortex line. Describe circles of radius vt with A and B as centres. If $vt < a$, the circles do not intersect; the disturbance is therefore wholly within them. But when vt is $> a$, the intersecting part contains no **H**, and only the **E** of the steady field due to the vortex line, which we know by § 24.

That within the part common to both circles there is no **H** we may prove thus. The vortex line in question may be imagined to be a line of latitude on any spherical surface passing through A and B, and centred upon the axis. Let a_1 be the radius of any sphere of this kind. Then, at a time making $vt > a$, the disturbance must lie between the surfaces of spheres of radii $vt-a_1$ and $vt+a_1$, whose centre is that of the sphere a_1. Now this excludes a portion of the space between the $vt-a$ and $vt+a$ circles, referring to the plane section; and by varying the radius a_1 we can find the whole space excluded. Thus, find the locus of intersections of circles of radius

$$vt-(a^2+z^2)^{\frac{1}{2}},$$

with centre at distance z from the origin, upon the axis. The equation of the circle is

$$(x-z)^2+y^2=\{vt-(a^2+z^2)^{\frac{1}{2}}\}^2,$$

or

$$x^2+y^2-2xz=v^2t^2+a^2-2vt(a^2+z^2)^{\frac{1}{2}}. \quad . \quad . \quad (172)$$

Differentiate with respect to z, giving

$$z(v^2t^2-x^2)^{\frac{1}{2}}=ax, \quad . \quad . \quad . \quad . \quad . \quad . \quad (173)$$

and eliminate z between (173) and (172). After reductions, the result is

$$x^2+(y\pm a)^2=v^2t^2, \quad . \quad . \quad . \quad . \quad . \quad . \quad (174)$$

indicating two circles, both of radius vt, whose centres are at A and B. Within the common space, therefore, the steady electric field has been established.

If this case be taken literally, then, since it involves an infinite concentration in a geometrical line of a finite amount of vorticity of **e**, the result for the steady field is infinite close up to that line, and the energy is infinite. But imagine, instead, the vorticity to be spread over a zone at the equator of the sphere $r=a$, half on each side of it, and its surface-density to be $f_1\nu$, where f_1 is finite. Consider the effect produced at a point in the equatorial plane. From time $t=0$ to $t_1=r-a$ (if the point be external) there is no disturbance. But from time t_1 to $t_2=b/v$, where b is the distance from the point to the edges of the zone, the disturbance must be identically the same as if the harmonic distribution $f_1\nu$ were complete, viz. by (142),

$$\mathrm{H}=\left(\frac{f_1\nu}{2\mu_0 v}\right)\frac{1}{2}\left(1+\frac{a^2-v^2t^2}{r^2}\right). \quad . \quad . \quad . \qquad (175)$$

After this moment t_2, the formula of course fails. Now narrow the band to width $ad\theta$ at the equator and simultaneously increase f_1, so as to make $f_1 ad\theta=e$, the strength of the shell of impressed force when there is but one. The formula (175) will now be true only for a very short time, and in the limit it will be true only momentarily, at the front of the wave, viz.

$$f_1 a/2\mu_0 vr=\mathrm{H}=e/2\mu_0 vr\, d\theta, \quad . \quad . \quad . \quad . \qquad (176)$$

going up infinitely as $d\theta$ is reduced. To avoid infinities in the electric and magnetic forces we must seemingly keep either to finite volume or finite surface-density of vorticity of **e**, just as in electrostatics with respect to electrification.

Instead of a simple shell of impressed electric force, it may be one of magnetic force, with similar results. As a verification calculate the displacement through circle ν on the sphere $r=a$ due to a vortex circle at ν_1 on the same surface, the latter being of unit strength. It is

$$\Sigma\frac{a\nu}{2}\frac{ce_m\nu Q'_m}{2m+1}, \quad . \quad . \quad . \quad . \quad . \qquad (177)$$

due to $\Sigma e_m Q_m$, through the circle ν. Take then

$$e_m=\frac{(2m+1)\nu_1{}^2 Q'_{1m}}{2m(m+1)}, \quad . \quad . \quad . \quad . \qquad (178)$$

which represents e_m due to vortex line of unit strength at ν_1. Use this in the preceding equation (177) and we obtain

$$\mathrm{D}=\Sigma\frac{ca}{4}\frac{\nu^2 Q'_m \nu^2{}_1 Q'_{1m}}{m(m+1)} \quad . \quad . \quad . \quad . \quad . \qquad (179)$$

as the displacement through ν due to unit vortex line at ν_1. Applying this result to a circular electric current, $\mathbf{B}=\mu_0\mathbf{H}$ takes the place of $\mathbf{D}=(c/4\pi)\mathbf{E}$, as the flux concerned, whilst if h be the strength of the shell of impressed magnetic force, $h/4\pi$ is the equivalent bounding electric current. The induction through the circle ν due to unit electric current in the circle ν_1 is therefore obtainable from (179) by turning c to μ_0 and multiplying by $(4\pi)^2$. The result agrees with Maxwell's formula for the coefficient of mutual induction of two circles (vol. ii. art. 697).

It must be noted that in the magnetic-shell application there must be no conductivity, if the wave-formulæ are to apply.

26. *An Electromotive Impulse.* $m=1$.—Returning to the case of impressed electric force, let in a spherical portion of an infinite dielectric a uniform field of impressed force act momentarily. We know the result of the continued application of the force. We have, then, to imagine it cancelled by an oppositely directed force, starting a little later. Let t_1 be the time of application of the real force, and let it be a small fraction of $2a/v$, the time the travelling shell takes to traverse any point. The result is evidently a shell of depth vt_1 at $r=vt+a$, in which the electromagnetic field is the same as in the case of continued application of the force, and a similar shell situated at $r=vt-a$, in which H is negative. Within this inner shell there is no **E** or **H**. But between the two thin shells just mentioned there is a diffused disturbance, of weak intensity, which is due to the sphericity of the waves, and would be non-existent were they plane waves. In fact, at time $t=t_1$, when the initial disturbance $\mathrm{H}=f_1\nu/2\mu_0 v$ has extended itself a small distance vt_1 on each side of the surface of the sphere, there is a radial component F at the surface itself, since, by (150),

$$\mathrm{F}_a = f_1 \cos\theta\left(\frac{vt}{a} - \frac{v^3t^3}{6a^3}\right), \quad . \quad . \quad . \quad (180)$$

so that the sudden removal of f_1 leaves two waves which do not satisfy the condition $\mathrm{E}=\mu_0 v\mathrm{H}$ at their common surface of contact. On separation, therefore, there must be a residual disturbance between them. The discontinuity in E at the moment of removing f_1 is abolished by instantaneous assumption of the mean value, but it is impossible to destroy the radial displacement which joins the two shells at the moment they separate. Put on f_1 when $t=0$, then $-f_1$ at time t_1 later. The H at time t due to both is by (142),

$$\mathrm{H}= \frac{f_1\nu}{4\mu_0 v r^2}v^2(t_1^2-2tt_1); \quad . \quad . \quad . \quad . \quad . \quad (181)$$

which, when t_1 is infinitely small, becomes

$$H = -\frac{f_1 \nu t t_1 v^2}{2\mu_0 v r^2}. \quad . \quad . \quad . \quad . \quad . \quad . \quad (182)$$

First of all, at a point distant r from the centre, comes the primary disturbance or head,

$$H = \frac{f_1 \nu a}{2\mu_0 v r}, \quad . \quad . \quad . \quad . \quad . \quad . \quad (183)$$

when $vt = r - a$, lasting for the time t_1. It is followed by the diffused negative disturbance, or tail, represented by (182), lasting for the time $2a/v$. At its end comes the companion to (182), its negative, when $vt = r + a$, lasting for time t_1, after which it is all over. This description applies when $r > a$. If $r < a$, the interval between the beginning and end of the H disturbance is only $2r/v$. From the above follows the integral solution expressing the effect of f_1 varying in any manner with the time.

27. *Alternating Impressed Forces.*—If the impressed force in the sphere, or wherever it may be, be a sinusoidal function of the time, making $p^2 = -n^2$, if $n = 2\pi \times$ frequency, the complete solutions arise from (132) to (135) so immediately that we can almost call them the complete solutions. Of course in any case in which we have developed the connexion between the impressed force and the flux, say $e = ZC$, or $C = Z^{-1}e$, when Z is the resistance operator, we may call this equation *the* solution in the sinusoidal case, if we state that p^2 is to mean $-n^2$. But there is usually a lot of work needed to bring the solution to a practical form. In the present instance, however, there is scarcely any required, because u and w are simple functions of qa, and q^2 is real. The substitution $p^2 = -n^2$ in u results in a real function of nr/v, and in w in a real function $\times (-1)^{\frac{1}{2}}$. Thus:—

$$\left.\begin{aligned} u_1 &= \cos\frac{nr}{v} - \frac{v}{nr}\sin\frac{nr}{v} \\ w_1 &= i\left(\sin\frac{nr}{v} + \frac{v}{nr}\cos\frac{nr}{v}\right) \end{aligned}\right\}, \quad . \quad . \quad . \quad . \quad (184)$$

$$\left.\begin{aligned} u_2 &= \left(1 - \frac{3v^2}{n^2r^2}\right)\cos\frac{nr}{v} - \frac{3v}{nr}\sin\frac{nr}{v} \\ w_2 &= i\left\{\left(1 - \frac{3v^2}{n^2r^2}\right)\sin\frac{nr}{v} + \frac{3v}{nr}\cos\frac{nr}{v}\right\} \end{aligned}\right\}. \quad . \quad (185)$$

In the case $m = 1$, if $(f_1)\cos nt$ is the form of f_1, so that

(f_1) represents the amplitude, we find, writing this case fully because it is the most important :—

$$\left.\begin{aligned}(\text{out})\quad \mathrm{H}&=\frac{(f_1)va}{\mu_0 vr}\left(\cos-\frac{v}{na}\sin\right)\frac{na}{v}.\left(\cos-\frac{v}{nr}\sin\right)\left(\frac{nr}{v}-nt\right)\\ (\text{in})\quad \mathrm{H}&=\frac{(f_1)va}{\mu_0 vr}\left(\cos-\frac{v}{nr}\sin\right)\frac{nr}{v}.\left(\cos-\frac{v}{na}\sin\right)\left(\frac{na}{v}-nt\right)\end{aligned}\right\}\quad(185\,a)$$

$$\left.\begin{aligned}(\text{out})\quad \mathrm{F}&=-\frac{2(f_1)va\mu}{nr^2}\left(\cos-\frac{v}{na}\sin\right)\frac{na}{v}.\left(\sin+\frac{v}{nr}\cos\right)\left(\frac{nr}{v}-nt\right)\\ (\text{in})\quad \mathrm{F}&=-\frac{2(f_1)va\mu}{nr^2}\left(\cos-\frac{v}{nr}\sin\right)\frac{nr}{v}.\left(\sin+\frac{v}{na}\cos\right)\left(\frac{na}{v}-nt\right)\end{aligned}\right\}\quad(185\,b)$$

$$\left.\begin{aligned}(\text{out})\quad \mathrm{E}&=\frac{(f_1)av}{r}\left(\cos-\frac{v}{na}\sin\right)\frac{na}{v}.\left\{\left(1-\frac{v^2}{n^2r^2}\right)\cos\right.\\ &\qquad\qquad\left.-\frac{v}{nr}\sin\right\}\left(\frac{nr}{v}-nt\right)\\ (\text{in})\quad \mathrm{E}&=-\frac{(f_1)av}{r}\left\{\left(1-\frac{v^2}{n^2r^2}\right)\sin+\frac{v}{nr}\cos\right\}\frac{nr}{v}.\left(\sin\right.\\ &\qquad\qquad\left.+\frac{v}{na}\cos\right)\left(\frac{na}{v}-nt\right)\end{aligned}\right\}\quad(185\,c)$$

It is very remarkable on first acquaintance that the impressed force produces no external effect at all when

$$u_a=0,\quad\text{or}\quad \tan\frac{na}{v}=\frac{na}{v}.$$

For the impressed force may be most simply taken to be a uniform field of intensity $(f_1)\cos nt$ in the sphere of radius a acting parallel to the axis, and it looks as if external displacement must be produced. Of course, on acquaintance with the reason, the fact that the solution is made up of two sets of waves, those outward from the lines of vorticity and those going inward, and then reflected out, the mystery disappears.

To show the positive and negative waves explicitly, we may write the first of (185 a) in the form

$$\begin{aligned}(\text{out})\quad \mathrm{H}=\frac{(f_1)av}{2\mu_0 vr}\Big[&\left\{\left(1-\frac{v^2}{n^2ar}\right)\cos+\left(\frac{v}{na}+\frac{v}{nr}\right)\sin\right\}\left(nt-\frac{n(a+r)}{v}\right)\\ &+\left\{\left(1+\frac{v^2}{n^2ar}\right)\cos+\left(\frac{v}{nr}-\frac{v}{na}\right)\sin\right\}\left(nt+\frac{n(a-r)}{v}\right)\Big],\end{aligned}\quad(185\,d)$$

the second line showing the primary wave out, the first the reflected wave *. Exchange a and r within the [ ] to obtain

* In reference to this formula (185 d), and the corresponding ones for other values of m, it is not without importance to know that a very

the internal H. The disturbance, at the surface, of the primary wave going both ways is, from $t=0$ to $2a/v$,

$$\frac{(f_1)\nu}{2\mu_0 v}\left\{\cos nt + \frac{v^2}{n^2 ar}(\cos nt - 1)\right\} \quad . \quad (185\,e)$$

The amplitude due to both waves is

$$\frac{(f_1)\nu}{\mu_0 v} u_a \left(1+\frac{v^2}{n^2 a^2}\right)^{\frac{1}{2}} \quad . \quad . \quad . \quad . \quad . \quad . \quad (185\,f)$$

The outward transfer of energy per second per unit area at any distance r is $\mathrm{EH}/4\pi$. In the mth system this is

$$\frac{\mathrm{EH}}{4\pi} = -\frac{(f_m)^2 a^2(\nu \mathrm{Q}_m')^2}{4\pi(\mu_0 v)^2 r} \cdot \frac{u_a^2}{cnr} \cdot \{u' \sin - (-iw')\cos\}\, nt \,.\, \{u \cos + (-iw)\sin\}\, nt, \quad (186)$$

where m is supposed odd, whilst u and $-iw$ are the real functions of nr/v obtained in the same way as (184). The mean value of the t function is, by the conjugate property of u and w, equation (114),

$$= -n/2v.$$

Using this, and integrating (186) over the complete surface of radius r, giving

$$\iint (\nu \mathrm{Q}_m')^2 d\mathrm{S} = \frac{4\pi r^2 m(m+1)}{2m+1}, \quad . \quad . \quad . \quad . \quad (187)$$

we find the mean transfer of energy outward per second through any surface enclosing the sphere to be

$$\frac{m(m+1)}{2(2m+1)} \frac{(f_m)^2 u_a^2 a^2}{\mu_0 v}, \quad . \quad . \quad . \quad . \quad . \quad (188)$$

if $(f_m)\nu \mathrm{Q}_m' \cos nt$ is the vorticity of the impressed force.

---

slight change suffices to make (185 d) represent the solution from the first moment of starting the impressed force. Thus, let it start when $t=0$, and let the f_1 in equation (185 d) be $(f_1)\cos nt$. Effect the two integrations thus,

$$\frac{f_1}{q} = (f_1)\frac{v}{n}\sin nt, \quad \frac{f_1}{q^2} = (f_1)\frac{v^2}{n^2}(1-\cos nt),$$

vanishing when $t=0$, and then operate with the exponentials, and we shall obtain (185 d) thus modified. To the first line must be added

$$\frac{(f_1)\nu a}{2\mu_0 vr}\,\frac{v^2}{n^2 ar},$$

and to the second line its negative. Thus modified, (185 d) is true from $t=0$, understanding that the second line begins when $t=(r-a)/v$, and the first when $t=(r+a)/v$. The first of (185 a) is therefore true up to distance $r=vt-2a$, when this is positive. In the shell of depth $2a$ beyond, it fails.

In the case $m=1$, the waste of energy per second is

$$\frac{(f_1)^2 a^2 u_a^2}{3\mu_0 v}, \quad . \quad . \quad . \quad . \quad . \quad . \quad . \quad (189)$$

due to the uniform alternating field of impressed force of intensity

$$(f_1) \cos nt$$

within the sphere.

In reality, the impressed force must have been an infinitely long time in operation to make the above solutions true to an infinite distance, and have therefore already wasted an infinite amount of energy. If the impressed force has been in operation any finite time t, however great, the disturbance has only reached the distance $r=vt+a$. Of course the solutions are true, provided we do not go further than $r=vt-a$. We see, therefore, that the real function of the never-ceasing waste of energy is to set up the sinusoidal state of **E** and **H** in the boundless regions of space to which the disturbances have not yet reached. The above outward waves are the same as in Rowland's solutions*. Here, however, they are explicitly expressed in terms of the impressed forces causing them.

$u_a=0$ makes the external field vanish when m is odd; and $w_a=0$ when m is even; that is, when the sinusoidal state has been assumed. It takes only the time $2a/v$ to do this, as regards the sphere $r=a$; the initial external disturbance goes out to infinity and is lost. This vanishing of the external field happens whatever may be the nature of the external medium away from the sphere, except that the initial external disturbance will behave differently, being variously reflected or absorbed according to circumstances.

28. *Conducting Medium.* $m=1$.—Now consider the same problem in an infinitely extended conductor of conductivity k. We may remark at once that, unless the conductivity is low, the solution is but little different from what it would be were the conductor not greatly larger than the spherical portion within it on whose surface lie the vortex lines of the impressed force, owing to the great attenuation suffered by the disturbances as they progress from the surface. In a similar manner, if the sphere be large, or the frequency of alternations great, or both, we may remove the greater part of the interior of the sphere without much altering matters.

We have now

$$q=(4\pi\mu_0 kp)^{\frac{1}{2}}=(1+i)x,$$

if

$$x=(2\pi\mu_0 kn)^{\frac{1}{2}}. \quad . \quad . \quad . \quad . \quad . \quad . \quad . \quad . \quad (190)$$

* In paper referred to in § 18.

The realization is a little troublesome on account of this $p^{\frac{1}{2}}$. The result is that the uniform alternating field of impressed force of intensity

$$(f_1)\cos nt,$$

gives rise to the internal solution

$$\text{(in)}\ \mathrm{H}=\left(\frac{\pi k}{2\mu_0 n}\right)^{\frac{1}{2}}\frac{(f_1)va}{r}\{(\mathrm{A}+\mathrm{B})\cos nt+(\mathrm{A}-\mathrm{B})\sin nt\}, \quad (191)$$

where A and B are the functions of r expressed by

$$\mathrm{A}=\epsilon^{x(r-a)}\left[\left(1+\frac{1}{2xa}-\frac{1}{2xr}\right)\cos+\left(\frac{1}{2xr}-\frac{1}{2xa}+\frac{2}{2xr\,.\,2xa}\right)\sin\right]x(a-r)$$

$$+\epsilon^{-x(r+a)}\left[\left(1+\frac{1}{2xa}+\frac{1}{2xr}\right)\cos-\left(\frac{1}{2xa}+\frac{1}{2xr}+\frac{2}{2xr\,.\,2xa}\right)\sin\right]x(a+r); \quad (192)$$

$$\mathrm{B}=\epsilon^{x(r-a)}\left[\left(\frac{1}{2xr}-\frac{1}{2xa}+\frac{2}{2xr\,.\,2xa}\right)\cos-\left(1-\frac{1}{2xr}+\frac{1}{2xa}\right)\sin\right]x(a-r)$$

$$-\epsilon^{-x(r+a)}\left[\left(\frac{1}{2xr}+\frac{1}{2xa}+\frac{2}{2xr\,.\,2xa}\right)\cos+\left(1+\frac{1}{2xr}+\frac{1}{2xa}\right)\sin\right]x(a+r). \quad (193)$$

Equation (191) showing the internal H, the external is got by exchanging a and r in the functions A and B.

Now xa is easily made large, in a good conductor ; then, anywhere near the boundary, $(r=a)$, we have

$$\left.\begin{aligned}\mathrm{A}&=\epsilon^{-x(a-r)}\cos x(a-r),\\ -\mathrm{B}&=\epsilon^{-x(a-r)}\sin x(a-r),\end{aligned}\right\}; \quad . \quad . \quad . \quad . \quad (194)$$

and (191) becomes

$$\text{(in)}\ \mathrm{H}=\left(\frac{\pi k}{\mu_0 n}\right)^{\frac{1}{2}}\frac{(f_1)av}{r}\epsilon^{-x(a-r)}.\cos\left\{nt-x(a-r)-\frac{\pi}{4}\right\}. \quad (195)$$

The wave-length λ is

$$\lambda=\left(\frac{2\pi}{\mu_0 kn}\right)^{\frac{1}{2}}. \quad . \quad . \quad . \quad . \quad . \quad . \quad (196)$$

Thus, in copper, a frequency of 1600 to 1700 makes $\lambda=$ 1 centim. Both λ and the attenuation-rate depend inversely on the square roots of the inductivity, conductivity, and frequency, whereas the amplitude varies directly as the square root of the conductivity, and inversely as the square roots of the others.

To verify that very great frequency ultimately limits the disturbance to the vortex line of $\mathbf{e}$ when there is but one, we may use the last solution to construct that due to a sheet of impressed force

$$\cos nt\,\Sigma\, e_m \mathrm{Q}_m$$

acting radially on the surface of the sphere. Thus,

$$\text{(in)}\ \mathrm{H}=\left(\frac{\pi k}{\mu_0 n}\right)^{\frac{1}{2}}\cdot\Sigma\frac{e_m \nu Q'_m}{r}\,\epsilon^{-x(a-r)}\cos\left\{nt-x(a-r)-\frac{\pi}{4}\right\} \quad (197)$$

when xa is very great. When the vorticity is confined to one line of latitude, H in (197) vanishes everywhere except at the vortex line. But a further approximation is required, or a different form of solution, to show the disturbance round the vortex line explicitly, *i. e.* when n is great, though not infinitely great.

29. *A Conducting Dielectric.* $m=1$.—Here, if k is the conductivity, c the permittivity, and μ_0 the inductivity, let

$$q=(4\pi\mu_0 kp+\mu_0 cp^2)^{\frac{1}{2}}=n_1+n_2 i, \quad . \quad . \quad . \quad (198)$$

when $p=ni$. Then n_1 and n_2 will be given by

$$\left.\begin{aligned} n_1^2&=\frac{n^2}{2v^2}\left[\left\{1+\left(\frac{4\pi k}{cn}\right)^2\right\}^{\frac{1}{2}}-1\right] \\ n_2^2&=\frac{n^2}{2v^2}\left[\left\{1+\left(\frac{4\pi k}{cn}\right)^2\right\}^{\frac{1}{2}}+1\right] \end{aligned}\right\} \quad . \quad . \quad (199)$$

Using this q in the general external H solution, but ignoring the explicit connexion with the impressed force, we shall arrive at

$$\text{(out)}\ \ \mathrm{H}=\frac{\mathrm{C}_0\nu}{r}\,\epsilon^{-n_1 r}\left[\left(1+\frac{n_1}{r(n_1^2+n_2^2)}\right)\cos - \frac{n_2}{r(n_1^2+n_2^2)}\sin\right](n_2 r-nt), \quad . \quad . \quad (200)$$

where C_0 is an undetermined constant, depending upon the magnitude of the disturbance at $r=a$. So far as the external solution goes, however, the internal connexions are quite arbitrary save in the periodicity and confinement to producing magnetic force proportional in intensity to the cosine of the latitude. The solution (200) may be continued unchanged as near to the centre as we please. Stopping it anywhere, there are various ways of constructing complementary distributions in the rest of space, from which (200) is excluded.

n_1 is zero when $k=0$. We then have the dielectric solution, with $n_2=n/v$. On the other hand, $c=0$ makes

$$n_1=n_2=(2\pi\mu_0 kn)^{\frac{1}{2}}=x,$$

as in § 28. The value of $n_1^2+n_2^2$ is

$$\frac{n^2}{v^2}\left(1+\left(\frac{4\pi k}{cn}\right)^2\right)^{\frac{1}{2}}=\frac{n^2}{v^2}\left(1+\left(\frac{4\pi k\mu_0 v^2}{n}\right)^2\right)^{\frac{1}{2}}. \quad . \quad . \quad (201)$$

Enormously great frequency brings us to the formulæ of the non-conducting dielectric, with a difference, thus: n_1 and n_2 become

$$n_1 = 2\pi k \mu_0 v, \quad n_2 = n/v, \quad . \quad . \quad . \quad . \quad (202)$$

when $4\pi k/cn$ is a small fraction. The attenuation due to conductivity still exists, but is independent of the frequency. We have now

$$\text{(out)} \quad \mathrm{H} = \frac{\mathrm{C}_0 \nu}{r} \epsilon^{-n_1 r} \left(\cos - \frac{v}{nr} \sin\right)\left(\frac{nr}{v} - nt\right), \quad (203)$$

differing from the case of no conductivity only in the presence of the exponential factor.

It is, however, easily seen by the form of n_1 in (202) that in a good conductor the attenuation in a short distance is very great, so that the disturbances are practically confined to the vortex lines of the impressed force, where the H disturbance is nearly the same as if the conductivity were zero, as before concluded. It follows that the initial effect of the sudden introduction of a steady impressed force in the conducting dielectric is the emission from the seat of its vorticity of waves in the same manner as if there were no conductivity, but attenuated at their front to an extent represented by the factor $\epsilon^{-n_1 r}$, with the (202) value of n_1, in addition to the attenuation by spreading which would occur were the medium non-conducting. This estimation of attenuation applies at the front only.

30. *Current in Sphere constrained to be uniform.*—Let us complete the solution (200) of § 29 by means of a current of uniform density parallel to the axis within the sphere of radius a, beyond which (200) is to be the solution. This will require a special distribution of impressed force, which we shall find. Equation (200) gives us the normal component of electric current at $r=a$, by differentiation. Let this be $\Gamma \cos \theta$. Then Γ is the density of the internal current. The corresponding magnetic field must have the boundary-value according to (200), and vary in intensity as the distance from the axis, its lines being circles centred upon it, and in planes perpendicular to it. Thus the internal **H** is also known. The internal **E** is fully known too, being $k^{-1}\Gamma$ in intensity and parallel to the axis. It only remains to find **e** to satisfy

$$\text{curl}\,(\mathbf{e} - \mathbf{E}) = \mu \dot{\mathbf{H}}, \quad . \quad . \quad . \quad . \quad . \quad (3)\ bis$$

within the sphere, and at its boundary (with the suitable surface interpretation), as it is already satisfied outside the sphere. The simplest way appears to be to first introduce a uniform field of **e** parallel to the axis, of such intensity e_1 as to neutralize the difference between the tangential components of the internal and external **E** at the boundary, and so make continuity there in the forc of the field; and next, to find an auxiliary distribution $\mathbf{e}_2$, such that

$$\text{curl}\,\mathbf{e}_2=\mu\dot{\mathbf{H}},$$

and having no tangential component on the boundary. This may be done by having $\mathbf{e}_2$ parallel to the axis, of intensity proportional to

$$(a^2-r^2)\sin\theta.$$

The result is that the internal H is got from the external by putting $r=a$ in (200) and then multiplying by r/a; Γ from the internal H by multiplying by $(2\pi r\sin\theta)^{-1}$; $\mathbf{e}_1$ from the difference of the tangential components E outside and inside is given by

$$e_1=\frac{C_0}{4\pi a^2}\epsilon^{-n_1a}\left\{k^2+\left(\frac{cn}{4\pi}\right)^2\right\}^{-1}\left[\left\{k\left(3+n_1a+\frac{3n_1}{a(n_1^2+n_2^2)}\right)\right.\right.$$
$$\left.+\frac{cn}{4\pi}\left(n_2a-\frac{3n_2}{a(n_1^2+n_2^2)}\right)\right\}\cos(n_2r-nt)$$
$$-\left\{k\left(\frac{3n_2}{a(n_1^2+n_2^2)}-n_2a\right)\right.$$
$$\left.\left.+\frac{cn}{4\pi}\left(3+n_1a+\frac{3n_1}{a(n_1^2+n_2^2)}\right)\right\}\sin(n_2r-nt)\right].\quad(204)$$

Finally, the auxiliary force has its intensity given by

$$e_2=\mu_0nC_0\frac{a^2-r^2}{2a^2}\epsilon^{-n_1a}\left\{\left(1+\frac{n_1}{a(n_1^2+n_2^2)}\right)\sin\right.$$
$$\left.+\frac{n_2}{a(n_1^2+n_2^2)}\cos\right\}(n_2a-nt).\quad(205)$$

A remarkable property of this auxiliary force, which (or an equivalent) is absolutely required to keep the current straight, is that it does no work on the current, on the average; the mean activity and waste of energy being therefore settled by e_1.

From the PHILOSOPHICAL MAGAZINE for October 1888.

ON ELECTROMAGNETIC WAVES, ESPECIALLY IN RELATION TO THE VORTICITY OF THE IMPRESSED FORCES; AND THE FORCED VIBRATIONS OF ELECTROMAGNETIC SYSTEMS.

BY

OLIVER HEAVISIDE.

On Electromagnetic Waves, especially in relation to the Vorticity of the Impressed Forces; and the Forced Vibrations of Electromagnetic Systems. By OLIVER HEAVISIDE.

[Continued from vol. xxv. p. 405.]

31. SPHERICAL *Waves (with diffusion) in a Conducting Dielectric.*—In an infinitely extended homogeneous isotropic conducting dielectric, let the surface $r=a$ be a sheet of vorticity of impressed electric force ; for simplicity, let it be of the first order, so that the surface-density is represented by $f\nu$. By (127), § 20, the differential equation of H, the intensity of magnetic force is, at distance r from the origin, outside the surface of f (ν meaning $\sin\theta$),

$$H=\frac{k_1}{q}\left(\frac{\nu a}{r}\right)\epsilon^{-qr}\left(1+\frac{1}{qr}\right)\left\{\cosh qa-\frac{\sinh qa}{qa}\right\}f, \quad . \quad (206)$$

where f may be any function of the time. Here, in the general case, including the unreal "magnetic conductivity" g,* we have

$$\left.\begin{aligned}q=[(4\pi k+cp)(4\pi g+\mu p)]^{\frac{1}{2}}=v^{-1}[(p+\rho)^2-\sigma^2]^{\frac{1}{2}},\\ k_1=4\pi k+cp\,;\end{aligned}\right\} \quad . \quad (207)$$

if, for subsequent convenience,

$$\left.\begin{aligned}\rho_1=4\pi k/2c,\quad \rho_2=4\pi g/2\mu,\quad v=(\mu c)^{-\frac{1}{2}}\,;\\ \rho=\rho_1+\rho_2,\quad \sigma=\rho_1-\rho_2.\end{aligned}\right\} \quad . \quad (208)$$

The speed is v, and ρ_1, ρ_2 are the coefficients of attenuation of the parts transmitted of elementary disturbances due to the real electric conductivity k and the unreal g; that is, $\epsilon^{-\rho t}$ is the factor of attenuation due to conductivity. On the

* Owing to the lapse of time, I should mention that the physical and other meanings of the coefficient g are explained in the first part of this Paper, Phil. Mag. Feb. 1888. Also k = electric conductivity; μ = magnetic inductivity; and $c/4\pi$ = electric permittivity. All the problems in this paper, except in § 43, relate to spherical waves; the geometrical coordinates are r and θ. Unless otherwise mentioned, p always signifies the operator d/dt, t being the time.

other hand, the distortion produced by conductivity depends on σ, and vanishes with it. There is some utility in keeping in g, because it sometimes happens that the vanishing of k, making $\rho = -\sigma$, leads to a solvable case. We can then produce a real problem by changing the meaning of the symbols, turning the magnetic into an electric field, with other changes to correspond.

32. *The steady Magnetic Field due to f constant.*—Let f be zero before, and constant after $t=0$, the whole medium having been previously free from electric and magnetic force. All subsequent disturbances are entirely due to f. The steady field which finally results is expressed by (206), by taking $p=0$; that is, k_1 has to mean $4\pi k$, and $q=4\pi(kg)^{\frac{1}{2}}$, by (207). To obtain the corresponding internal field, exchange a and r in (206), except in the first a/r. The same values of k_1 and q used in the corresponding equations of E and F give the final electric field. The steady magnetic field here considered depends upon g, and vanishes with it.

33. *Variable state when $\rho_1 = \rho_2$. First case. Subsiding f.*—There are cases in which we already know how the final state is reached, viz. the already given case of a nonconducting dielectric (§§ 21, 22), and the case $\sigma=0$ in (208), which is an example of the theory of § 4. In the latter case the impressed force must subside at the same rate as do the disturbances it sends out from the surface of f. Thus, given $f=f_0\epsilon^{-\rho t}$, starting when $t=0$, with f_0 constant, the resulting electric and magnetic fields are represented by those in the corresponding case in a nonconducting dielectric, when multiplied by $\epsilon^{-\rho t}$. The final state is zero because f subsides to zero; the travelling shell also loses all its energy. But there are, in a sense, two final states; the first commencing at any place as soon as the rear of the travelling shell reaches it, and which is entirely an electric field; the second is zero, produced by the subsidence of this electric field. There is no magnetic field to correspond, and therefore no "true" electric current, in Maxwell's sense of the term, except in the shell.

34. *Second case. f constant.*—But let the impressed f be constant. Then, by effecting the integrations in (206), we are immediately led to the full solution

$$\mathrm{H} = \frac{fva}{2\mu vr}\left[\epsilon^{-\frac{\rho}{v}(r-a)}\left(1+\frac{v}{\rho r}\right)\left(1-\frac{v}{\rho a}\right)+\epsilon^{-\rho t}(1+\rho t)\frac{v^2}{ra\rho^2}\right.$$
$$\left. + \text{same function of } -a\right], \quad . \quad . \quad (209)$$

where the fully represented part expresses the primary wave out from the surface of f, reaching r at time $(r-a)/v$; whilst the rest expresses the second wave, reaching r when $t=(r+a)/v$. After that the actual H is their sum, viz.

$$H=\frac{fva}{\mu vr}\epsilon^{-\rho r/v}\left(1+\frac{v}{\rho r}\right)\left[\cosh-\frac{v}{\rho a}\sinh\right]\frac{\rho a}{v}, \quad . \quad (210)$$

agreeing with (206), when we give q therein the special value ρ/v at present concerned, and $k_1=4\pi k$.

At the front of the first wave we have

$$H=\epsilon^{-\rho t}fva/2\mu vr, \quad . \quad . \quad . \quad . \quad . \quad (211)$$

so that the energy in the travelling shell still subsides to zero. Equation (211) also expresses H at the front of the inward wave, both before and after reaching the centre of the sphere. The exchange of a and r in the [] in (209) produces the corresponding internal solution.

35. *Unequal* ρ_1 *and* ρ_2. *General case.*—If we put $d/dr=\nabla$, we may write (206) thus,

$$H=\frac{va}{2r}\frac{k_1}{q^3}\left[\left(\nabla-\frac{1}{r}\right)\left(\nabla+\frac{1}{a}\right)\epsilon^{-q(r-a)}+\left(\nabla-\frac{1}{r}\right)\left(\nabla-\frac{1}{a}\right)\epsilon^{-q(r+a)}\right]f. \quad (212)$$

It is, therefore, sufficient to find

$$\epsilon^{-q(r-a)}q^{-3}f, \quad . \quad . \quad . \quad . \quad . \quad . \quad (213)$$

to obtain the complete solution of (212); namely, by performing upon the solution (213) the differentiations ∇ and the operation k_1. This refers to the first half of (212); the second half only requires the sign of a to be changed in the [].

Now (213) is the same as

$$v^3\epsilon^{-\rho t}\epsilon^{-\frac{r-a}{v}(p^2-\sigma^2)^{\frac{1}{2}}}(p^2-\sigma^2)^{-\frac{3}{2}}(f\epsilon^{\rho t}). \quad . \quad . \quad . \quad (214)$$

Expand the two functions of p in descending powers of p, thus,

$$(p^2-\sigma^2)^{-\frac{3}{2}}=p^{-3}\left[1+\frac{3}{2}\frac{\sigma^2}{p^2}+\frac{3.5}{2^2\lfloor 2}\frac{\sigma^4}{p^4}+\frac{3.5.7}{2^3\lfloor 3}\frac{\sigma^6}{p^6}+\ldots\right], \quad (215)$$

$$\epsilon^{-\frac{r-a}{v}(p^2-\sigma^2)^{\frac{1}{2}}}=\epsilon^{-\frac{p}{v}(r-a)}\left[1+\frac{\sigma}{p}h_1+\frac{\sigma^2}{p^2}h_2+\ldots\right], \quad . \quad . \quad . \quad (216)$$

where the h's are functions of r, but not of p. Multiplying these together, we convert (213) or (214) to

$$v^3\epsilon^{-\rho t}\epsilon^{-\frac{p}{v}(r-a)}\frac{1}{p^3}\left[1+\frac{\sigma}{p}i_1+\frac{\sigma^2}{p^2}i_2+\ldots\right](f\epsilon^{\rho t}), \quad . \quad (217)$$

where the i's are functions of r, but not of p. The integrations can now be effected. Let f be constant, first. Then

f starting when $t=0$, we have

$$p^{-3}(f\epsilon^{\rho t})=f\rho^{-3}(\epsilon^{\rho t}-1-\rho t-\tfrac{1}{2}\rho^2t^2)=\rho^{-3}f(\epsilon^{\rho t})_3 \text{ say}; \quad . \quad (218)$$

&c. &c. Next, operating with the exponential containing p in (217) turns t to $t-(r-a)/v$, and gives the required solution in the form

$$\mathrm{H}=\frac{fvav^2}{2\mu vr}\epsilon^{-\rho t}(\sigma+p)\Big[\Big(\nabla-\frac{1}{r}\Big)\Big(\nabla+\frac{1}{a}\Big)\Big\{\frac{(\epsilon^{\rho t_1})_3}{\rho^3}+\sigma i_1\frac{(\epsilon^{\rho t_1})_4}{\rho^4}+\dots\Big\}$$
$$+ \text{ same function of } -a\Big], \quad . \quad (219)$$

where $t_1=t-(r-a)/v$; the represented part beginning when t_1 reaches zero, and the rest when $t-(r+a)/v$ reaches zero.

36. *Fuller development in a special case. Theorems involving Irrational Operators.*—As this process is very complex, and (219) does not admit of being brought to a readily interpretable form, we should seek for special cases which are, when fully developed, of a comparatively simple nature. Write the first half of (212) thus,

$$\mathrm{H}=\frac{cv^3va}{2r}\epsilon^{-\rho t}\Big(\nabla-\frac{1}{r}\Big)\Big(\nabla+\frac{1}{a}\Big)\frac{1}{p^2-\sigma^2}\Big[\Big(\frac{p+\sigma}{p-\sigma}\Big)^{\frac{1}{2}}\epsilon^{-\frac{r-a}{v}(p^2-\sigma^2)^{\frac{1}{2}}}(f\epsilon^{\rho t})\Big]. (220)$$

Now the part in the square brackets can be finitely integrated when $f\epsilon^{\rho t}$ subsides in a certain way. We can show that

$$\Big(\frac{p+\sigma}{p-\sigma}\Big)^{\frac{1}{2}}\epsilon^{-\frac{r-a}{v}(p^2-\sigma^2)^{\frac{1}{2}}}(\epsilon^{-\sigma t})=\mathrm{J}_0\Big\{\frac{\sigma}{v}[(r-a)^2-v^2t^2]^{\frac{1}{2}}\Big\}, \quad (221)$$

in which, observe, the sign of σ may be changed, making no difference on the right side (the result), but a great deal on the left side.

The simplest proof of (221) is perhaps this. First let $r=a$. Then

$$\Big(\frac{p+\sigma}{p-\sigma}\Big)^{\frac{1}{2}}(\epsilon^{-\sigma t})=\epsilon^{-\sigma t}\Big(1-\frac{2\sigma}{p}\Big)^{-\frac{1}{2}}(1), \quad . \quad . \quad (222)$$

by getting the exponential to the left side, so as to operate on unity. Next, by the binomial theorem,

$$=\epsilon^{-\sigma t}\Big[1+\frac{1}{2}\frac{2\sigma}{p}+\frac{1\cdot 3}{2^2\lfloor 2}\Big(\frac{2\sigma}{p}\Big)^2+\dots\Big](1). \quad . \quad (223)$$

Now integrate, and we have (f commencing when $t=0$),

$$\left.\begin{aligned}&=\epsilon^{-\sigma t}\Big(1+\sigma t+\frac{1\cdot 3}{\lfloor 2\lfloor 2}\sigma^2t^2+\frac{1\cdot 3\cdot 5}{\lfloor 3\lfloor 3}\sigma^3t^3+\dots\Big),\\ &=\epsilon^{-\sigma t}\epsilon^{\sigma t}\mathrm{J}_0(\sigma ti);\end{aligned}\right\} \quad . (224)$$

so that, finally,

$$\left(\frac{p+\sigma}{p-\sigma}\right)^{\frac{1}{2}}(\epsilon^{-\sigma t}) = J_0(\sigma t i). \quad . \quad . \qquad (225)$$

It is also worth notice that, integrating in a similar manner,

$$\left(1-\frac{\sigma^2}{p^2}\right)^{-\frac{1}{2}}(1) = 1+\tfrac{1}{2}\frac{\sigma^2 t^2}{\underline{|2}} + \frac{1\,.\,3}{2^2\underline{|2}}\frac{\sigma^4 t^4}{\underline{|4}} + \ldots,$$

$$= J_0(\sigma t i). \quad . \quad . \quad . \quad . \quad . \quad . \quad . \quad . \qquad (226)$$

These theorems present themselves naturally in problems relating to a telegraph-circuit, when treated by the method of resistance-operators. A special case of (225) is

$$p^{\frac{1}{2}}(1) = (\pi t)^{-\frac{1}{2}}, \quad . \quad . \quad . \quad . \quad . \qquad (227)$$

which presents itself in the electrostatic theory of a submarine* cable.

We have now to generalize (225) to meet the case (221). The left member of (221) satisfies the partial differential equation

$$v^2\nabla^2 = p^2 - \sigma^2, \quad . \quad . \quad . \quad . \quad . \quad . \quad . \qquad (228)$$

so we have to find the solution of (228) which becomes $J_0(\sigma t i)$ when $r=a$. Physical considerations show that it must be an even function of $(r-a)$, so that it is suggested that the t in J_0 $(\sigma t i)$ has to become, not $t-(r-a)/v$ or $t+(r-a)/v$, but that t^2 has to become their product. In any case, the right member of (221) does satisfy (228) and the further prescribed condition, so that (221) is correct.

* Thus, let an infinitely long circuit, with constants R, S, K, L, be operated upon by impressed force at the place $z=0$, producing the potential-difference V_0 there, which may be any function of the time. Let C be the current and V the potential-difference at time t at distance x. Then

$$C = \left(\frac{K+Sp}{R+Lp}\right)^{\frac{1}{2}}\epsilon^{-qz}V_0,$$

where $q=(R+Lp)^{\frac{1}{2}}(K+Sp)^{\frac{1}{2}}$. Take $K=0$, and $L=0$; then, if V_0 be zero before and constant after $t=0$, the current at $z=0$ is given by

$$C_0 = V_0(S/R)^{\frac{1}{2}}p^{\frac{1}{2}}(1),$$

and (227) gives the solution. Prove thus: let b be any constant, to be finally made infinite; then

$$p^{\frac{1}{2}}(1) = b^{\frac{1}{2}}(1+bp^{-1})^{-\frac{1}{2}}$$
$$= b^{\frac{1}{2}}J_0(\tfrac{1}{2}bti)\epsilon^{-bt/2}$$

by the investigation in the text. Now put $b=\infty$, and (227) results.

In the similar treatment of cylindrical waves in a conductor, $p^{\frac{1}{4}}$, $p^{\frac{3}{4}}$, &c. occur. We may express these results in terms of Gamma functions.

If a direct proof be required, expand the exponential operator in (221) containing r in the way indicated in (216), and let the result operate upon $J_0(\sigma t i)$. The integrated result can be simplified down to (221).

37. Now use (221) in (220). Let $f\epsilon^{pt}=f_0\epsilon^{-\sigma t}$, where f_0 is constant; and the square bracket in (220) becomes known, being in fact the right member of (221) multiplied by f_0. So, making use also of (228), we bring (220) to

$$H=\frac{vaf_0}{2\mu vr}\epsilon^{-pt}\left[1+\left\{\left(\frac{1}{a}-\frac{1}{r}\right)v^2\frac{d}{dr}-\frac{v^2}{ar}\right\}\frac{1}{p^2-\sigma^2}\right]$$
$$J_0\left\{\frac{\sigma}{v}\left[(r-a)^2-v^2t^2\right]^{\frac{1}{2}}\right\}; \quad (229)$$

to which must be added the other part, beginning $2a/v$ later, got by negativing a, except the first one. The operation $(p^2-\sigma^2)^{-1}$ may be replaced by two integrations with respect to r.

Let r and a be infinitely great, thus abolishing the curvature. Let $r-a=z$, and f_0va/r, which is now constant, be called e_0. Then we have simply

$$H=\frac{e_0}{2\mu v}\epsilon^{-pt}J_0\left\{\frac{\sigma}{v}(z^2-v^2t^2)^{\frac{1}{2}}\right\}, \quad . \quad . \quad . \quad (230)$$

showing the H produced in an infinite homogeneous conducting dielectric medium at time t after the introduction of a plane sheet (at $z=0$), of vorticity of impressed electric force, the surface density of vorticity being $e_0\epsilon^{-2\rho_1 t}$. This corroborates the solution in § 8, equation (51) (vol. xxv. p. 140), whilst somewhat extending its meaning.

The condition to which f is subject may be written, by (208),

$$f=f_0\epsilon^{-2\rho_1 t}, \quad . \quad . \quad . \quad . \quad . \quad . \quad (231)$$

where f_0 is constant. If, then, we desire f to be constant, ρ_1 must vanish, which, by (208), requires $k=0$, whilst g may be finite.

But we can make the problem real thus. In (229) change H to E and μv to cv; we have now the solution of the problem of finding the electric field produced by suddenly magnetizing uniformly a spherical portion of a conducting dielectric; *i. e.* the vorticity of the impressed magnetic force is to be on the surface of the sphere $r=a$, parallel to its lines of latitude, and of surface-density fv, such that $fv\epsilon^{2\rho_2 t}$ is constant. This makes f constant when $g=0$ and k finite, representing a real conducting dielectric.

38. *The electric force at the origin due to fv at $r=a$.*—Returning to the case of impressed electric force, the differential equation of F, the radial component of electric force inside the sphere on whose surface $r=a$ the vorticity of **e** is situated, is, by § 2D, equations (136), (137),

$$\mathrm{F}=\frac{2a\cos\theta}{qr^2}\left(1+\frac{1}{qa}\right)\left(\cosh qr-\frac{\sinh qr}{qr}\right)f, \quad (232)$$

At the centre, therefore, the intensity of the full force, which call F_0, whose direction is parallel to the axis, is

$$\mathrm{F}_0=\tfrac{2}{3}(1+qa)\epsilon^{-qa}f=\tfrac{2}{3}\left(1-a\frac{d}{da}\right)\epsilon^{-qa}f. \quad . \quad . \quad (233)$$

Unless otherwise specified, I may repeat that the forces referred to are always those of the fluxes, thus doing away with any consideration of the distribution of the impressed force, and of scalar potential, of varying form, which it involves. (233) is equivalent to

$$\mathrm{F}_0=\tfrac{2}{3}\,\epsilon^{-pt}\{1+av^{-1}(p^2-\sigma^2)^{\frac{1}{2}}\}\epsilon^{-av^{-1}(p^2-\sigma^2)^{\frac{1}{2}}}(f\epsilon^{pt}) \,. \quad (234)$$

Let f be constant, and $\rho=\sigma$, or $g=0$. Then (234) becomes

$$\mathrm{F}_0=\tfrac{2}{3}f\epsilon^{-\sigma t}\left[\frac{a}{v}(p+\sigma)\left(\frac{p-\sigma}{p+\sigma}\right)^{\frac{1}{2}}+1\right]\epsilon^{-av^{-1}(p^2-\sigma^2)^{\frac{1}{2}}}(\epsilon^{\sigma t}), \quad . \quad (235)$$

of which the complete solution is, by (221),

$$\mathrm{F}_0=(\tfrac{2}{3}f)[\epsilon^{-\sigma t}av^{-1}(p+\sigma)\mathrm{J}_0\{\sigma v^{-1}(a^2-v^2t^2)^{\frac{1}{2}}\}+\mathrm{X}_a], \quad (236)$$

where, subject to $g=0$,

$$\epsilon^{-qa}(1)=\mathrm{X}_a\,; \quad . \quad . \quad . \quad . \quad . \quad . \quad (236a)$$

or, solved,

$$\mathrm{X}_a=1-\epsilon^{-\sigma t}\left[\frac{\sigma a}{v}\left(\mathrm{J}_0+\frac{\mathrm{J}_1}{i}\right)-\frac{1}{\underline{|3}}\left(\frac{\sigma a}{v}\right)^3\frac{1}{\sigma t}\left(\frac{\mathrm{J}_1}{i}+\frac{\mathrm{J}_2}{i^2}\right)\right.$$

$$\left.+\frac{1\,.\,3}{\underline{|5}}\left(\frac{\sigma a}{v}\right)^5\frac{1}{\sigma^2t^2}\left(\frac{\mathrm{J}_2}{i^2}+\frac{\mathrm{J}_3}{i^3}\right)-\ldots\right], \quad (237)$$

in which $i=(-1)^{\frac{1}{2}}$, and all the J's operate upon σti. This solution (236) begins when $t=a/v$. The value of σ is $4\pi k/2c$.

In a good conductor σ is immense. Then assume $c=0$, or do away with the elastic displacement, and reduce (236) to the pure diffusion formula, which is

$$\mathrm{F}_0=(\tfrac{2}{3}f)\left[\left(\frac{2}{\pi}\right)^{\frac{1}{2}}y\epsilon^{-\frac{1}{2}y^2}+1-\left(\frac{2}{\pi}\right)^{\frac{1}{2}}\left\{y-\frac{y^3}{\underline{|3}}+\frac{1\,.\,3}{\underline{|5}}y^5-\ldots\right\}\right], \quad (238)$$

where $y=(4\pi\mu ka^2/2t)^{\frac{1}{2}}$. The relation of X_a in (236) to the preceding terms is explained by equations (233) or (235).

39. *Effect of uniformly magnetizing a Conducting Sphere surrounded by a Nonconducting Dielectric.*—Here, of course, it is the lines of **E** that are circles centred upon the axis, both inside and outside. Let **h** be the impressed magnetic force, and $h\nu$ the surface-density of its vorticity, at $r=a$, outside which the medium is nonconducting, and inside a conducting dielectric. The differential equation of E_a, the surface value of the tensor of **E** at $r=a$, is [compare (124), § 19]

$$\frac{h\nu}{E_a}=\left(\frac{1}{\mu p}\frac{W'}{W}\right)_{out}-\left(\frac{1}{\mu p}\frac{U'+W'}{U+W}\right)_{in}; \quad . \quad . \quad . \quad (239)$$

in which $r=a$, and μ and q are to have the proper values on the two sides of the surface.

Now, by (111),

$$W'/W=-q\{1+(qr)^{-1}(1+qr)^{-1}\} \quad . \quad . \quad . \quad (240)$$

in the case of $m=1$, (first order), here considered. This refers to the external dielectric, in which $q=p/v$. Let $v=\infty$, making

$$W'/W=-a^{-1}. \quad . \quad . \quad . \quad . \quad . \quad (241)$$

This assumption is justifiable when the sphere has sensible conductivity, on account of the slowness of action it creates in comparison with the rapidity of propagation in the dielectric outside. Then (239) becomes,

$$-\frac{h\nu}{E_a}=\frac{1}{\mu_1 pa}\frac{q_1 a \sinh q_1 a}{\cosh q_1 a-(q_1 a)^{-1}\sinh q_1 a}+\frac{1}{pa}\left(\frac{1}{\mu_0}-\frac{1}{\mu_1}\right), (242)$$

if μ_0 is the external and μ_1 the internal inductivity, and q_1 the internal q. When the inductivities are equal, there is a material simplification, leading to

$$E_a=-\mu pa\frac{\cosh q_1 a-(q_1 a)^{-1}\sinh q_1 a}{q_1 a \sinh q_1 a}h\nu, \quad . \quad (243)$$

where $q_1=\{(4\pi k_1+c_1 p)\mu_1 p\}^{\frac{1}{2}}$. First let $c_1=0$, in the conductor, making $q_1^2=4\pi\mu_1 k_1 p=-s^2$, say. Then

$$E_a=-\frac{1}{4\pi k_1 a}\frac{\cos sa-(sa)^{-1}\sin sa}{(sa)^{-1}\sin sa}h\nu. \quad . \quad . \quad (244)$$

From this we see that $\sin sa=0$ is the determinantal equation of normal systems. The slowest is

$$sa=\pi, \text{ or } -p^{-1}=4\mu_1 k_1 a^2/\pi. \quad . \quad . \quad . \quad (245)$$

This time-constant is about $(1250)^{-1}$ second if the sphere be of copper of 1 centim. radius; about 8 seconds if of 1 metre radius, and about 10 million years if of the size of the earth.

At distance r from the centre of the sphere, within it, at time t after starting h, we have

$$\mathrm{E} = -\frac{hv}{4\pi k_1 r}\Sigma\frac{\cos sr-(sr)^{-1}\sin sr}{p(d/dp)\{(sa)^{-1}\sin sa\}}\frac{\epsilon^{pt}}{\cos sa}, \quad . \quad . \quad (246)$$

subject to the determinantal equation, over whose roots the summation extends, p being now algebraic. Effecting the differentiation indicated, we obtain

$$\mathrm{E} = -\frac{2hv}{4\pi k_1 r}\Sigma\frac{\cos sr-(sr)^{-1}\sin sr}{\cos sa}\epsilon^{pt}. \quad . \quad . \quad (247)$$

The corresponding solution for the radial component of the magnetic force, say H_r, is

$$\mathrm{H}_r = (\tfrac{2}{3}h\cos\theta)-4h\cos\theta\,\Sigma\frac{\cos sr-(sr)^{-1}\sin sr}{s^2r^2\cos sa}\epsilon^{pt}. \quad . \quad (248)$$

At the centre of the sphere, let H_0 be the intensity of the actual magnetic force. It is, by (248),

$$\mathrm{H}_0 = \tfrac{2}{3}h\{1+2\,\Sigma\,(\cos sa)^{-1}\epsilon^{pt}\}. \quad . \quad . \quad (249)$$

Thus the magnetic force arrives at the centre of the sphere in identically the same manner as current arrives at the distant end of an Atlantic cable according to the electrostatic theory, when a steady impressed force is applied at the beginning, with terminal short-circuits. In the case of the cable the first time-constant is

$$-p^{-1} = \mathrm{RS}l^2/\pi^2,$$

where $\mathrm{R}l$ is the total resistance and $\mathrm{S}l$ the total permittance. It is not greatly different from 1 second, so that, by (245), the sphere should be about a foot in radius to imitate, at its centre, the arrival curve of the cable.

To be precise we should not speak of magnetizing the sphere, because (ignoring the minute diamagnetism) it does not become magnetized. The principle, however, is the same. We set up the flux magnetic induction. But the magnetic terminology is defective. Perhaps it would be not objected to if we say we inductize* the sphere, whether we magnetize it or not. This is, at any rate, better than extending the meaning of the word magnetize, which is already precise in the mathematical theory, though of uncertain application in practice, from the variable behaviour of iron.

* Accent the first syllable, like magnetize. Practical men sometimes speak of energizing a core, &c. But energize is too general; by using inductize we specify what flux is set up.

40. The following is the alternative form of solution showing the waves, when c_1 is finite. With the same assumption as before that $v=\infty$ outside the sphere, the equation of H_r the radial component of $\mathbf{H}$ is

$$H_r=\frac{2\cos\theta}{q^2r^2}\frac{\cosh qr-(qr)^{-1}\sinh qr}{(qa)^{-1}\sinh qa}h, \quad . \quad . \quad (250)$$

which, at $r=0$, becomes

$$H_0=\tfrac{2}{3}qa(\sinh qa)^{-1}h. \quad . \quad . \quad . \quad . \quad (251)$$

Expand the circular function, giving

$$H_0=\tfrac{4}{3}qa\,\epsilon^{-qa}(1+\epsilon^{-2qa}+\epsilon^{-4qa}+\ldots)h\,; \quad . \quad (252)$$

or, since here $q=v^{-1}\{(p+\sigma)^2-\sigma^2\}^{\frac{1}{2}}$,

$$H_0=\frac{4}{3}\frac{a}{v}\,\epsilon^{-\sigma t}(p+\sigma)\left(\frac{p-\sigma}{p+\sigma}\right)^{\frac{1}{2}}\left[\epsilon^{-\frac{a}{v}(p^2-\sigma^2)^{\frac{1}{2}}}+\epsilon^{-\frac{3a}{v}(p^2-\sigma^2)^{\frac{1}{2}}}+\ldots\right](h\epsilon^{\sigma t}), \quad . \quad (253)$$

so, using (221), we get finally

$$H_0=\frac{4}{3}h\frac{a}{v}\epsilon^{-\sigma t}(p+\sigma)\left[J_0\left\{\frac{\sigma}{v}(a^2-v^2t^2)^{\frac{1}{2}}\right\}+J_0\left\{\frac{\sigma}{v}(9a^2-v^2t^2)^{\frac{1}{2}}\right\}+\ldots\right]. \quad (254)$$

The J_0 functions commence when $vt=a$, $3a$, $5a$, &c., in succession, and the successive terms express the arrival of the first wave and of the reflexions from the surface which follow. In the case of pure diffusion, this reduces to

$$H_0=(\tfrac{2}{3}h)2a(4\pi k_1\mu_1/\pi t)^{\frac{1}{2}}[\epsilon^{-\pi k_1\mu_1a^2/t}+\epsilon^{-9\pi\mu_1k_1a^2/4t}+\ldots], \quad (255)$$

which is the alternative form of (249), involving instantaneous action at a distance. The theorem (in diffusion)

$$\epsilon^{-xp^{\frac{1}{2}}}.p^{\frac{1}{2}}(1)=(\pi t)^{-\frac{1}{2}}\epsilon^{-x^2/4t} \quad . \quad . \quad . \quad (256)$$

becomes generalized to

$$\epsilon^{-xq}q(1)=v^{-1}\epsilon^{-\sigma t}(p+\sigma)J_0\{\sigma v^{-1}(x^2-v^2t^2)^{\frac{1}{2}}\}, \quad . \quad (257)$$

if

$$q=v^{-1}(p^2+2\sigma p)^{\frac{1}{2}}.$$

On the right side of (257), the p means, as usual, differentiation to t. The two quantities σ and v may have any positive values; to reduce to (256) make v infinite whilst keeping σ/v^2 finite.

41. *Diffusion of Waves from a centre of impressed force in a conducting medium.*—In equation (206) let a be infinitely small. It then becomes

$$H=\tfrac{1}{3}a^3\nu r^{-2}(4\pi k+cp)(1+qr)\epsilon^{-qr}f, \quad . \quad . \quad (258)$$

the equation of H at distance r from an element of impressed electric force at the origin. Comparing with (233) we see that the solution of (258) may be derived, when f is constant, starting when $t=0$. Take $g=0$, making $\rho=\sigma=4\pi k/2c$. Then

$$\mathrm{H}=\frac{\mathrm{H}_0}{4\pi k}(4\pi k+cp)\left\{\epsilon^{-\sigma t}\frac{r}{v}(p+\sigma)\mathrm{J}_0\left[\frac{\sigma}{v}(r^2-v^2t^2)^{\frac{1}{2}}\right]+\mathrm{X}_r\right\}, \quad (259)$$

where X_r is what the X_a of (237) becomes on changing a to r; and

$$\mathrm{H}_0=\nu kr^{-2}\times\text{vol. integral of } f, \quad . \quad . \quad . \quad (260)$$

supposing the impressed force to be confined to the infinitely small sphere, so that its volume-integral is the "electric moment," by analogy with magnetism. The solution (259) begins at r as soon as $t=r/v$. It is true from infinitely near the origin to infinitely near the front; but no account is given of the state of things at the front itself. H_0 is the final value of H. We may also write X_r thus,

$$\epsilon^{\sigma t}\mathrm{X}_r=1+\int_r^{vt}\frac{dr}{v}(\sigma+p)\mathrm{J}_0\left\{\frac{\sigma}{v}(r^2-v^2t^2)^{\frac{1}{2}}\right\}; \quad . \quad (261)$$

and (259) may also be written

$$\mathrm{H}=\frac{\mathrm{H}_0}{4\pi k}(4\pi k+cp)\left(1-r\frac{d}{dr}\right)\mathrm{X}_r. \quad . \quad . \quad (262)$$

When $c=0$, (259) or (262) reduce to

$$\mathrm{H}=\mathrm{H}_0\left[\left(\frac{2}{\pi}\right)^{\frac{1}{2}}y\epsilon^{-\frac{1}{2}y^2}+1-\left(\frac{2}{\pi}\right)^{\frac{1}{2}}\left\{y-\frac{1}{\underline{|3}}y^3+\frac{1.3}{\underline{|5}}y^5-\ldots\right\}\right], \quad (263)$$

where $y=(2\pi\mu kr^2/t)^{\frac{1}{2}}$.

42. *Conducting sphere in nonconducting dielectric. Circular vorticity of* **e**. *Complex reflexion. Special very simple case.*—At distance r from the origin, outside the sphere of radius a, which is the seat of vorticity of **e**, represented by fv, we have

$$\mathrm{H}=\phi^{-1}(\mathrm{W}/\mathrm{W}_a)fva/r. \quad . \quad . \quad . \quad (264)$$

The operator ϕ will vary according to the nature of things on both sides of $r=a$. When it is a uniform conducting medium inside, and nonconducting outside, to infinity, we shall have

$$\phi=\phi_1+\phi_2,$$

where ϕ_1, depending upon the inner medium, is given by

$$\phi_1=\frac{q_1}{4\pi k_1+c_1p}\,\frac{\{1+(q_1a)^{-2}\}\sinh q_1a-(q_1a)^{-1}\cosh q_1a}{\cosh q_1a-(q_1a)^{-1}\sinh q_1a}, \quad (265)$$

and ϕ_2, depending upon the outer medium, is given by

$$\phi_2 = \mu v \frac{1+(qa)^{-1}+(qa)^{-2}}{1+(qa)^{-1}}. \quad . \quad . \quad . \quad (266)$$

The solution arising from the sudden starting of f constant is therefore

$$H = (fva/r)\Sigma(W/W_a)(p \,.\, d\phi/dp)^{-1}\epsilon^{pt}, \quad . \quad (267)$$

where p is now algebraical, and the summation ranges over the roots of $\phi=0$. There is no final H in this case, if we assume $g=0$ all over. But the determinantal equation is very complex, so that this (267) solution is not capable of easy interpretation. The wave method is also impracticable, for a similar reason.

In accordance, however, with Maxwell's theory of the impermeability of a "perfect" conductor to magnetic induction from external causes, the assumption $k_1=\infty$ makes the solution depend only upon the dielectric, modified by the action of the boundary, and an extraordinary simplification results. ϕ_1 vanishes, and the determinantal equation becomes $\phi_2=0$, which has just two roots,

$$qa = pa/v = -\tfrac{1}{2} + i\left(\frac{3}{4}\right)^{\frac{1}{2}}; \quad . \quad . \quad . \quad . \quad (268)$$

and these, used in (267), give us the solution

$$H = (fva/3\mu vr)\epsilon^{-z}\{3\cos - 3^{\frac{1}{2}}(1-2a/r)\sin\}z\sqrt{3}, \quad . \quad (269)$$

where $z = \{vt-(r-a)\}/2a$.

Correspondingly, the tangential and radial components of **E** are

$$E = \mu v H + fva^3r^{-3}[1-\tfrac{1}{3}\epsilon^{-z}(3\cos + 3^{\frac{1}{2}}\sin)z\sqrt{3}] \quad . \quad . \quad (270)$$

$$F = \frac{2a^3}{r^3} f\cos\theta \left[1 - \frac{1}{3}\frac{r}{a}\epsilon^{-z}\left(\frac{3a}{r}\cos - \sqrt{3}\left(2-\frac{a}{r}\right)\sin\right)z\sqrt{3}\right]. \quad (271)$$

This remarkably simple solution, considering that there is reflexion, corroborates Prof. J. J. Thomson's investigation* of the oscillatory discharge of an infinitely conducting spherical shell initially charged to surface-density proportional to the sine of the latitude, for, of course, it does not matter how thin or thick the shell may be when infinitely conducting, so that it may be a solid sphere. (269) to (271) show the establishment of the permanent state. Take off the impressed force,

* "On Electrical Oscillations and the Effects produced by the Motion of an Electrified Sphere," Proc. Math. Soc. vol. xv. p. 210.

and the oscillatory discharge follows. But the impressed force keeping up the charge on the sphere need not be an external cause, as supposed in the paper referred to. There seems no other way of doing it than by having impressed force with vorticity fv on the surface, but in other respects it is immaterial whether it is internal or external, or superficial.

It may perhaps be questioned whether the sphere does reflect, seeing that its surface is the seat of f. But we have only to shift the seat of f to an outer spherical surface in the dielectric, to see at once that the surface of the conductor is the place of continuous reflexion of the wave incident upon it coming from the surface of f. The reflexion is not, however, of the same simple character that occurs when a plane wave strikes a plane boundary ($k=\infty$) flush, which consists merely in sending back again every element of **H** unchanged, but with its **E** reversed; the curvature makes it much more complex. When we bring the surface of f right up to the conducting sphere, we make the reflexion instantaneous. At the front of the wave we have $z=0$ and

$$\mathrm{H}=fva/\mu vr=\mathrm{E}/\mu v$$

by (269) and (270). This is exactly double what it would be were the conductor replaced by dielectric of the same kind as outside, the doubling being due to the instantaneous reflexion of the inward going wave by the conductor.

The other method of solution may also be applied, but is rather more difficult. We have

$$\mathrm{H}=\frac{av}{\mu vr}\epsilon^{-q(r-a)}\left(1+\frac{1}{qr}\right)\left(1-\frac{1}{qa}\right)\left(1-\frac{1}{q^3a^3}\right)^{-1}f. \quad . \quad (272)$$

Expand the last factor in descending powers of $(qa)^3$, and integrate. The result may be written

$$\mathrm{H}=\frac{afv}{\mu vr}\left(\frac{d^2}{dx^2}+\left(\frac{a}{r}-1\right)\left(\frac{d}{dx}-\frac{a}{r}\right)\left(\frac{x^2}{\underline{|2}}+\frac{x^5}{\underline{|5}}+\frac{x^8}{\underline{|8}}+\ldots\right), (273)$$

where $x=a^{-1}(vt-r+a)$. Conversion to circular functions reproduces (269).

42 A. *Same case with finite Conductivity. Sinusoidal Solution.*—It is to be expected that with finite conductivity, even with the greatest at command, or $k=(1600)^{-1}$, the solution will be considerably altered, being controlled by what now happens in the conducting sphere. To examine this point, consider only the value of H at the boundary. We have, by (264),

$$\mathrm{H}_a=\phi^{-1}fv=(\phi_1+\phi_2)^{-1}fv. \quad . \quad . \quad . \quad (274)$$

Let f vary sinusoidally with the time, and observe the behaviour of ϕ_1 and ϕ_2 as the frequency changes. The full development which I have worked out is very complex. But it is sufficient to consider the case in which k is big enough, in concert with the radius a and frequency $n/2\pi$, to make the disturbances in the sphere be practically confined to a spherical shell whose depth is a small part of the radius. Let $s=(2\pi\mu_1 k_1 na^2)^{\frac{1}{2}}$; then our assumption requires ϵ^{-s} to be small. This makes

$$\phi_1=-\frac{1}{4\pi k_1 a}\left(1-2s^2\frac{s+i(s-1)}{1-2s+2s^2}\right), \quad . \quad . \quad (275)$$

and, if further, s itself be a large number, this reduces to

$$\phi_1=(1+i)\ (\mu_1 n/8\pi k_1)^{\frac{1}{2}}. \quad . \quad . \quad . \quad . \quad (276)$$

Adding on the other part of ϕ, similarly transformed by $p^2=-n^2$, we obtain

$$\phi=\left(\mu v\frac{(na/v)^2}{1+(na/v)^2}+\left(\frac{\mu_1 n}{8\pi k_1}\right)^{\frac{1}{2}}\right)-i\left(\frac{\mu v}{(na/v)+(na/v)^3}-\left(\frac{\mu_1 n}{8\pi k_1}\right)^{\frac{1}{2}}\right), (277)$$

where the terms containing k_1 show the difference made by its not being infinite. The real part is very materially affected. Thus, copper, let

$$k_1=(1600)^{-1},\ \mu_1=1,\ 2\pi n=1600,\ a=10,\ \therefore\ s=10.$$

These make s large enough. Now na/v is very small, but, on the other hand,

$$(\mu_1 n/8\pi k_1)^{\frac{1}{2}}=130,$$

so that the real part of ϕ depends almost entirely on the sphere, whilst the other part is little affected.

Now make n extremely great, say $na/v=1$; else the same. Then

$$\phi=(\tfrac{3}{2}\times 10^{10}+44\times 10^4)-i(\tfrac{3}{2}\times 10^{10}-44\times 10^4),$$

from which we see that the dissipation in space has become *relatively* important. The ultimate form, at infinite frequency, is

$$\phi=\mu v+(\mu_1 n/8\pi k_1)^{\frac{1}{2}}(1+i); \quad . \quad . \quad . \quad . \quad (278)$$

so that we come to a third state, in which the conductor puts a stop to all disturbance. This is, however, because it has been assumed not to be a dielectric also, so that inertia ultimately controls matters. But if, as is infinitely more probable, it is a dielectric, the case is quite changed. We shall have

$$\phi_1=(4\pi g_1+\mu_1 p)^{\frac{1}{2}}(4\pi k_1+c_1 p)^{-\frac{1}{2}}, \quad . \quad . \quad . \quad (279)$$

when the frequency is great enough, and this tends to $\mu_1 v_1$, μ_1 being the inductivity and v_1 the speed in the conductor, whatever g and k may be, provided they are finite. Thus, finally,

$$\phi = \mu_1 v_1 + \mu v \quad . \quad . \quad . \quad . \quad . \quad . \quad (280)$$

represents the impedance, or ratio of $f v$ to H_a, which are now in the same phase.

At any distance outside we know the result by the dielectric solution for an outward wave. But there is only superficial disturbance in the conducting sphere.

43. *Resistance at the front of a wave sent along a wire.*—In its entirety this question is one of considerable difficulty, for two reasons, if not three. First, although we may, for practical purposes, when we send a wave along a telegraph-circuit, regard it as a plane wave, in the dielectric, on account of the great length of even the short waves of telephony, and the great speed, causing the lateral distribution (out from the circuit) of the electric and magnetic fields to be, to a great distance, almost rigidly connected with the current in the wires and the charges upon them; yet this method of representation must to some extent fail at the very front of the wave. Secondly, we have the fact that the penetration of the electromagnetic field into the wires is not instantaneous; this becomes of importance at the front of the wave, even in the case of a thin wire, on account of the great speed with which it travels over the wire *. The resistance per unit length must vary rapidly at the front, being much greater there than in the body of the wave; thus causing a throwing back, equivalent to electrostatic or "jar" retardation.

Now, according to the electromagnetic theory, the resistance must be infinitely great at the front. Thus, alternate the current sufficiently slowly, and the resistance is practically the steady resistance. Do it more rapidly, and produce appreciable departure from uniformity of distribution of current in the wire, and we increase the resistance to an amount calculable by a rather complex formula. But do it

* The distance within which, reckoned from the front of the wave backward, there is material increased resistance, we may get a rough idea of by the distance travelled by the wave in the time reckoned to bring the current-density at the axis of the wire to, say, nine tenths of the final value. It has all sorts of values. It may be 1 or 1000 kilometres, according to the size of wire and material. At the front, on the assumption of constant resistance, the attenuation is according to $\epsilon^{-Rt/2L}$, R being the resistance, and L the inductance of the circuit per unit length. Hence the importance of the increased resistance in the present question.

very rapidly, and cause the current to be practically confined to near the boundary, and we have a simplified state of things in which the resistance varies inversely as the area of the boundary, which may, in fact, be regarded as plane. The resistance now increases as the square root of the frequency, and must therefore, as said, be infinitely great at the front of a wave, which is also clear from the fact that penetration is only just commencing.

But for many reasons, some already mentioned, it is far more probable that the wire is a dielectric. If, as all physicists believe, the æther permeates all solids, it is *certain* that it is a dielectric. Now this becomes of importance in the very case now in question, though of scarcely any moment otherwise. Instead of running up infinitely, the resistance per unit area of surface of a wire tends to the finite value $4\pi\mu_1 v_1$. This is great, but far from infinity, so that the attenuation and change of shape of wave at its front produced by the throwing back cannot be so great as might otherwise be expected.

Thus, in general, at such a great frequency that conduction is nearly superficial, we have, if μ, c, k, and g belong to the wire,

$$E/H = (4\pi g + \mu p)^{\frac{1}{2}}(4\pi k + cp)^{-\frac{1}{2}}, \quad . \quad . \quad . \quad (281)$$

if E is the tangential electric force and H the magnetic force, also tangential, at the boundary of a wire. Now let R′ and L′ be the resistance and inductance of the wire per unit of its length. We must divide H by 4π to get the corresponding current in the wire, as ordinarily reckoned. So $4\pi A^{-1}$ times the right number of (281) is the resistance-operator of unit length, if A is the surface per unit length; so, expanding (281), we get

$$R' \text{ or } L'n = \frac{4\pi}{A}\frac{\mu v}{\sqrt{2}}\left\{\left(\frac{4\rho_2^2 + n^2}{4\rho_1^2 + n^2}\right)^{\frac{1}{2}} \pm \frac{4\rho_1\rho_2 + n^2}{4\rho_1^2 + n^2}\right\}^{\frac{1}{2}}, \quad (282)$$

where ρ_1, ρ_2 are as before, in (208). Here $n/2\pi =$ frequency.

Disregarding g, and ρ_2, we have

$$R' \text{ or } L'n = (\tfrac{1}{2})^{\frac{1}{2}} 4\pi\mu v A^{-1}\{B \pm B^2\}^{\frac{1}{2}}, \quad . \quad . \quad . \quad (283)$$

where

$$B = n(4\rho_1^2 + n^2)^{-\frac{1}{2}} = nc\{(4\pi k)^2 + n^2c^2\}^{-\frac{1}{2}}.$$

When c is zero, R′ and L′n tend to equality, as shown by Lord Rayleigh. But when c is finite, L′n tends to *zero*, and R′ to $4\pi\mu v A^{-1}$, as indeed we can see from (281) at once, by the relative evanescence of k and g, when finite.

But the frequency needed to bring about an approximation towards the constant resistance is excessive; in copper, we require trillions per second. This brings us to the third reason mentioned ; we have no knowledge of the properties of matter under such circumstances, or of æther either. The net result is that although it is infinitely more probable that the resistance should tend to constancy than to infinity, yet the real value is quite speculative*. Similar remarks apply to sudden discharges, as of lightning along a conductor. The above R, it should be remarked, is real resistance, in spite of its ultimate form, suggestive of impedance without resistance†. The present results are corroborative of those in Part I., and, in fact, only amount to a special application of the same.

44. *Reflecting Barriers.*—Let the medium be homogeneous between $r=a_0$ and $r=a_1$, where there is a change of some kind, yet unstated. Let between them the surface $r=a$ be a sheet of vorticity of **e** of the first order. We already know what will happen when fv is started, for a certain time, until in fact the inward wave reaches the inner boundary, and, on the other side, until the outward wave reaches the outer boundary; though, unless the surface of f is midway between the boundaries, the reflected wave from the nearest barrier will reach into a portion of the region beyond f, by the time the further barrier is reached by the primary wave. The

* The above was written before the publication of Professor Lodge's highly interesting lectures before the Society of Arts. Some of the experiments described in his second lecture are seemingly quite at variance with the electromagnetic theory. I refer to the smaller impedance of a short circuit of fine iron wire than of thick copper, as reckoned by the potential-difference at its beginning needed to spark across the circuit between knobs. Should this be thoroughly verified, it has occurred to me as a possible explanation that things may be sometimes so nicely balanced that the occurrence of a discharge may be determined by the state of the skin of the wire. A wire cannot be homogeneous right up to its boundary, with then a perfectly abrupt transition to air; and the electrical properties of the transition-layer are unknown. In particular, the skin of an iron wire may be nearly unmagnetizable, μ varying from 1 to its full value, in the transition-layer. Consequently, in the above formula, resistance $4\pi\mu v$ per unit surface, we may have to take $\mu=1$ in the extreme, in the case of an iron wire. But even then, the explanation of Professor Lodge's results is capable of considerable elucidation. Perhaps resonance will do it.

† There is a tendency at present amongst some writers to greatly extend the meaning of resistance in electromagnetism; to make it signify cause/effect. This seems a pity, owing to the meaning of resistance having been thoroughly specialized in electromagnetism already, in strict relationship to "frictional" dissipation of energy. What the popular meaning of "resistance" may be is beside the point. I would suggest that what is now called the magnetic resistance be called the magnetic reluctance; and per unit volume, the reluctancy.

subsequent history depends upon the constitution of the media beyond the boundaries, which can be summarized in two boundary conditions. The expression for E/H is, in general,

$$\frac{\mathrm{E}}{\mathrm{H}}=-(4\pi k+cp)^{-1}\frac{u'-yw'}{u-yw}, \quad . \quad . \quad . \quad (284)$$

by (120), extended, the extension being the introduction of y, which is a differential operator of unstated form, depending upon the boundary conditions. Let y_0 and y_1 be the y's on the inner and outer side of the surface of f. The differential equations of H_a, the magnetic force there, is then

$$fv=\{(\mathrm{E}/\mathrm{H})_{(\mathrm{out})}-(\mathrm{E}/\mathrm{H})_{(\mathrm{in})}\}\,\mathrm{H}_a, \quad . \quad . \quad (285)$$

as in § 19. Applying (284) and the conjugate property (114) of the functions u and w, (since there is no change of medium at the surface of f), this becomes

$$\mathrm{H}_a=\frac{4\pi k+cp}{q}\,\frac{(u_a-y_0w_a)(u_a-y_1w_a)}{y_1-y_0}fv; \quad . \quad (286)$$

from which the differential equation of H at any point between a_0 and a is obtained by changing $u_a-y_0w_a$ to $(a/r)(u-y_0w)$; and at any point between a and a_1 by changing $u_a-y_1w_a$ to $(a/r)(u-y_1w)$.

Unless, therefore, there are singularities causing failure, the determinantal equation is

$$y_1-y_0=0, \quad . \quad . \quad . \quad . \quad . \quad . \quad . \quad (287)$$

and the complete solution between a_0 and a_1 due to f constant may be written down at once. Thus at a point outside the surface of f we have

$$(\mathrm{out})\ \ \mathrm{H}=\frac{4\pi k+cp}{q}\,\frac{a}{r}\,\frac{(u_a-y_0w_a)(u-y_1w)}{y_1-y_0}fv=\phi^{-1}f; \quad (288)$$

and therefore, if f starts when $t=0$,

$$\mathrm{H}=\frac{f}{\phi_0}+\frac{fav}{r}\Sigma\frac{(u_a-yw_a)(u-yw)}{p(d/dp)(y_1-y_0)}\cdot\frac{4\pi k+cp}{q}\epsilon^{pt}, \quad . \quad (289)$$

p being now algebraic, given by (287); ϕ_0 the steady ϕ, from (288); and y the common value of the (now) equal y's; which identity makes (289) applicable on both sides of the surface of f.

45. *Construction of y_1 and y_0.*—In order that y_1 and y_0 should be determinable in such a way as to render (286) true, the media beyond the boundaries must be made up of any

number of concentric shells, each being homogeneous, and having special values of c, k, μ, and g. For the spherical functions would not be suitable otherwise, except during the passage of the primary waves to the boundaries, or until they reached places where the departure from the assumed constitution commenced. Assuming the constitution in homogeneous spherical layers, there is no difficulty in building up the forms of y_0 and y_1 in a very simple and systematic manner, wholly free from obscurities and redundancies. In any layer the form of E/H is as in (284), containing one y. Now at the boundary of two layers E is continuous, and also H (provided the physical constants are not infinite), so E/H is continuous. Equating, therefore, the expressions for E/H in two contiguous media expresses the y of one in terms of the y of the other. Carrying out this process from the origin up to the medium between a_0 and a, expresses y_0 in terms of the y of the medium containing the origin; this is zero, so that y_0 is found as an explicit function of the values of u, w, u', w' at all the boundaries between the origin and a_0. In a similar manner, since the y of the outermost region, extending to infinity, is 1, we express y_1, belonging to the region between a and a_1, in terms of the values of u, &c. at all the boundaries between a and ∞. Each of these four functions will occur twice for each boundary, having different values of the physical constants with the same value of r. I mention this method of equation of E/H operators because it is a far simpler process than what we are led to if we use the vector and scalar potentials; for then the force of the flux has three component vectors—the impressed force, the slope of the scalar potential, and the time-rate of decrease of the vector potential. The work is then so complex that a most accomplished mathematician may easily go wrong over the boundary conditions. These remarks are not confined in application to spherical waves.

If an infinite value be given to a physical constant, special forms of boundary condition arise, usually greatly simplified; *e. g.* infinite conductivity in one of the layers prevents electromagnetic disturbances from penetrating into it from without; so that they are reflected without loss of energy.

Knowing y_1 and y_0 in (288), we virtually possess the sinusoidal solution for forced vibrations, though the initial effects, which may or may not subside or be dissipated, will require further investigation for their determination; also the solution in the form of an infinite series showing the effect of suddenly starting f constant; also the solution arising from any initial distribution of **E** and **H** of the kind appropriate to the functions, viz. such as may be produced by vorticity of **e** in spherical layers, proportional to ν (or $\nu Q'_m$ in general). But it is scarcely necessary to say that these solutions in infinite series, of so very general a character, are more ornamental than useful. On the other hand, the immediate integration of the differential equations to show the development of waves becomes excessively difficult, from the great complexity, when there is a change of medium to produce reflexion.

46. *Thin Metal Screens.*—This case is sufficiently simple to be useful. Let there be at $r=a_1$ a thin metal sheet interposed between the inner and outer nonconducting dielectrics, the latter extending to infinity. If made infinitely thin, E is continuous, and H discontinuous to an amount equal to 4π times the conduction current (tangential) in the sheet. Let K_1 be the conductance of the sheet (tangential) per unit area; then

$$(H/E)_{in}-(H/E)_{out}=4\pi K_1 \text{ at } r=a_1.$$

Therefore by (284), when the dielectric is the same on both sides,

$$cp\left(\frac{u_1-w_1}{u_1'-w_1'}-\frac{u_1-y_1w_1}{u_1'-y_1w_1'}\right)=4\pi K_1,$$

where the functions u_1, &c. have the $r=a_1$ values. From this,

$$y_1=\frac{1-(4\pi K_1/cpq)u_1'(u_1'-w_1')}{1-(4\pi K_1/cpq)w_1'(u_1'-w_1')} \quad . \quad . \quad . \quad (290)$$

expresses y_1 for an outer thin conducting metal screen, to be used in (286). If of no conductivity, it has no effect at all, passing disturbances freely, and $y_1=1$. At the other extreme we have infinite conductivity, making $y_1=u_1'/w_1'$, with complete stoppage of outward-going waves, and reflexion without absorption, destroying the tangential electric disturbance.

When the screen, on the other hand, is within the surface of f, say, at $r=a_0$, of conductance K_0 per unit area, we shall find

$$y_0=\frac{(4\pi K_0/cpq)u_0'^2}{1+(4\pi K_0/cpq)u_0'w_0'}, \quad . \quad . \quad . \quad (291)$$

where u_0, &c. have the $r=a_0$ values. The difference of form from y_1 arises from the different nature of the r functions in the region including the origin. As before, no conductivity gives transparency ($y_0=0$), and infinite conductivity total reflexion ($y_0=u_0'/w_0'$). When the inner screen is shifted up to the origin, we make $y_0=0$ and so remove it.

47. *Solution with outer screen* ; $K_1=\infty$; f *constant.*—Let there be no inner screen and let the outer be perfectly conducting. As J. J. Thomson has considered these screens*, I will be very brief, regarding them here only in relation to the sheet of f and to former solutions. The determinantal equation is

$$u_1'=0, \text{ or } \tan x=x(1-x^2)^{-1}, \quad . \quad . \quad . \quad (292)$$

if $x=ipa/v$. Roots nearly π, 2π, 3π, &c.; except the first, which is considerably less. The solution due to starting f constant, by (289) is therefore

$$H=(fva/\mu vr)\Sigma(uu_a w_1'/a_1 u_1'')\epsilon^{pt}; \quad . \quad . \quad (293)$$

which, developed by pairing terms, leads to

$$H=\frac{fva}{\mu vr}\Sigma\frac{x^4-x^2+1}{x^3(x^2-2)}2\sin\frac{vtx}{a_1}\left(\cos-\frac{a_1}{xr}\sin\right)\frac{xr}{a_1}\left(\cos-\frac{a_1}{xa}\sin\right)\frac{xa}{a_1}, \quad ($$

which of course includes the effects of the infinite series of reflexions at the barrier. By making $a_1=\infty$, however, the result should be the same as if the screen were non-existent, because an infinite time must elapse before the first reflexion can begin, and we are concerned only with finite intervals. The result is

$$H=\frac{fva}{\mu vr}\cdot\frac{2}{\pi}\int_0^\infty dx_1\frac{\sin x_1 vt}{x_1}\left(\cos-\frac{1}{x_1 r}\sin\right)x_1 r\left(\cos-\frac{1}{x_1 a}\sin\right)x_1 a, \quad (29$$

which must be the equivalent of the simple solution (142) of § 21, showing the origin and progress of the wave.

Now reduce it to a plane wave. We must make a infinite, and $r-a=z$ finite. Also take $fv=e$, constant. We then have

$$H=\frac{e}{4\mu v}\frac{2}{\pi}\int_0^\infty dx_1\left(\sin\frac{x_1(vt-z)}{x_1}+\frac{\sin x_1(vt+z)}{x_1}\right), \quad . \quad (296)$$

showing the H at x due to a plane sheet of vorticity of e situated at $z=0$. This is the equivalent of the solution (12) of § 2, indicating the continuous uniform emission of $H=e/2\mu v$ both ways from the plane $z=0$.

Returning to (294), it is clear that from $t=0$ to $t=(a_1-a)/v$, the solution is the same as if there were no screen. Also if a is a very small fraction of a_1, the electromagnetic wave of depth $2a$ will, when it strikes the screen, be reflected nearly as from a plane boundary. It would therefore seem that this

* In the paper before referred to.

wave would run to and from between the origin and boundary unceasingly. This is to a great extent true; and therefore there is no truly permanent state (the electric flux, namely, alone); but examination shows that the reflexion is not clean, on account of the electrification of the boundary, so that there is a spreading of the magnetic field all over the region within the screen.

48. *Alternating f with reflecting barriers. Forced vibrations.* —Let the medium be nonconducting between the boundaries a_0 and a_1. Equation (288) then becomes

$$H = \frac{\nu a}{\mu v r} \frac{(u_a - y_0 w_a)(u - y_1 w)}{y_1 - y_0} f, \quad . \quad . \quad (297)$$

giving H outside the surface of f. We see that $y_0 = 0$ and $u_a = 0$ make $H = 0$. That is, the forced vibrations are confined to the inside of the surface of f only, at the frequencies given by $u_a = 0$, provided there is no internal screen to disturb, but independently of the structure of the external medium (since y_1 is undetermined so far), with possible exceptions due to the vanishing of y_1 simultaneously. But (297), sinusoidally realized by $p^2 = -n^2$, does not represent the full final solution, unless the nature of y_0 and y_1 is such as to allow the initial departure from this solution to be dissipated in space or killed by resistance. Ignoring the free vibrations, let $y_0 = 0$, and $y_1 = u_1'/w_1'$, meaning no internal, and an infinitely conducting external screen. Then

$$\left. \begin{array}{ll} \text{(out)} & H = (\nu a/\mu v r) u_a \{ u w_1'/u_1' - w \} f, \\ \text{(in)} & H = (\nu a/\mu v r) u \ \{ u_a w_1'/u_1' - w_a \} f. \end{array} \right\} \quad . \quad . \quad (298)$$

If $w_1' = 0$, or in full,

$$(v/na_1) \tan (na_1/v) = 1 - (v/na_1)^2,$$

we obtain a simplification, viz.

$$H_{(\text{in or out})} = -(\nu a/\mu v r)(u w_a \text{ or } u_a w) f ; \quad . \quad . \quad (299)$$

and the corresponding tangential components of electric force are

$$E_{(\text{in or out})} = (\nu a/\mu v r)(u' w_a \text{ or } u_a w')(cp)^{-1} f. \quad . \quad (300)$$

But if $u_1' = 0$, the result is infinite. This condition indicates that the frequency coincides with that of one of the free vibrations possible within the sphere $r = a_1$ without impressed force. But, considering that we may confine our impressed force to as small a space as we please round the origin, the infinite result is not easily understood, as regards its development.

But the development of infinitely great magnetic force by a *plane* sheet of f is very easily followed in full detail, not merely with sinusoidal f, but with f constant. Considering the latter case, the emission of H is continuous, as before described, from the surface of f. Now place a plane infinitely conducting barrier parallel to f, say on the left side. We at once stop the disturbances going to the left and send them back again, unchanged as regards **H**, reversed as regards **E**. The **H** disturbance on the left side of f therefore commences to be doubled after the time a/v has elapsed, a being the distance of the reflecting barrier from the plane of f, and on the right side after the interval $2a/v$. Next, put a second infinitely conducting barrier on the right side of f. It also doubles the **H** disturbances as they arrive; so that, by the inclusion of the plane of f between impermeable barriers, combined with the continuous emission of **H**, the magnetic disturbance mounts up infinitely, in a manner which may be graphically followed with ease. Similarly, with f alternating, at particular frequencies depending upon the distances of the two barriers from f.

Returning to the spherical case, an infinitely conducting internal screen, with no external, produces

$$H_a = \frac{(u_a w_0' - w_a u_0')(u_a - w_a)}{\mu v (w_0' - u_0')} fv. \quad . \quad . \quad . \quad (301)$$

We cannot produce infinite H in this case, because the absence of an external barrier will not let it accumulate. Shifting the surface of f right up to the screen, or *vice versâ*, simplifies matters greatly, reducing to the case of § 42.

May 8th, 1888.

From the PHILOSOPHICAL MAGAZINE for November 1888.

On Electromagnetic Waves, especially in relation to the Vorticity of the Impressed Forces; and the Forced Vibrations of Electromagnetic Systems. By OLIVER HEAVISIDE.

[Continued from p. 382.]

CYLINDRICAL ELECTROMAGNETIC WAVES.

49. IN concluding this paper I propose to give some cases of cylindrical waves. They are selected with a view to the avoidance of mere mathematical developments and unin-

telligible solutions, which may be multiplied to any extent; and for the illustration of peculiarities of a striking character. The case of vibratory impressed E.M.F. in a thin tube is very rich in this respect, as will be seen later. At present I may remark that the results of this paper have little application in telegraphy or telephony, when we are only concerned with long waves. Short waves are, or may be, now in question, demanding a somewhat different treatment*. We do, however, have very short waves in the discharge of condensers, and in vacuum-tube experiments, so that we are not so wholly removed from practice as at first appears. But independently of considerations of practical realization, I am strongly of opinion that the study of very unrealizable problems may be of use in forwarding the supply of one of the pressing wants of the present time or near future, a practicable æther—mechanically, electromagnetically, and perhaps also gravitationally comprehensive.

50. *Mathematical Preliminary.*—On account of some peculiarities in Bessel's functions, which require us to change the form of our equations to suit circumstances, it is desirable to exhibit separately the purely mathematical part. This will also considerably shorten and clarify what follows it.

Let the axis of z be the axis of symmetry, and let r be the distance of any point from it. Either the lines of **E**, electric force, or of **H**, magnetic force, may be circular, centred on the axis. For definiteness, choose **H** here. Then the lines of **E** are either longitudinal, or parallel to the axis; or there is, in addition, a radial component of **E**, parallel to r. Thus the tensor H of **H**, and the two components of **E**, say E longitudinal and F radial, fully specify the field. Their connexions are these special forms of equations (2) and (3):—

* The waves here to be considered are essentially of the same nature as those considered by J. J. Thomson, "On Electrical Oscillations in a Cylindrical Conductor," Proc. Math. Soc. vol. xvii., and in Parts I. and II. of my paper "On the Self-Induction of Wires," Phil. Mag. August and September 1886; viz. a mixture of the plane and cylindrical. But the peculiarities of the telegraphic problem make it practically a case of plane waves as regards the dielectric, and cylindrical in the wires. The "resonance" effects described in my just-mentioned paper arise from the to-and-fro reflexion of the plane waves in the dielectric, moving parallel to the wire. This is also practically true in Prof. Lodge's recent experiments, discharging a Leyden jar into a miniature telegraph-circuit. On the other hand, most of such effects in the present paper depend upon the cylindrical waves in the dielectric; and, in order to allow the dielectric fair play for their development, the contaminating influence of diffusion is done away with by using tubes only when there are conductors. In Hertz's recent experiments the waves are of a very mixed character indeed.

$$\left.\begin{aligned} \frac{1}{r}\frac{d}{dr}r\mathrm{H}=(4\pi k+cp)\mathrm{E}, \quad -\frac{d\mathrm{H}}{dz}=(4\pi k+cp)\mathrm{F}, \\ \frac{d\mathrm{E}}{dr}-\frac{d\mathrm{F}}{dz}=\mu p\mathrm{H}, \end{aligned}\right\} \quad (302)$$

where (and always later) p stands for d/dt. This is in space where neither the impressed electric nor the impressed magnetic force has curl, it being understood that **E** and **H** are the forces of the fluxes, so as to include impressed. From (302) we obtain

$$\left.\begin{aligned} \frac{1}{r}\frac{d}{dr}r\frac{d\mathrm{E}}{dr}+\frac{d^2\mathrm{E}}{dz^2}=(4\pi k+cp)\mu p\mathrm{E}, \\ \frac{d}{dr}\frac{1}{r}\frac{d}{dr}r\mathrm{H}+\frac{d^2\mathrm{H}}{dz^2}=(4\pi k+cp)\mu p\mathrm{H}, \end{aligned}\right\} \quad . \quad (303)$$

the characteristics of E and H. Let now

$$q^2=-s^2=(4\pi k+cp)\mu p-d^2/dz^2; \quad . \quad . \quad (304)$$

then the first of (303) becomes the equation of $\mathrm{J}_0(sr)$ and its companion, whilst the second becomes that of $\mathrm{J}_1(sr)$, and its companion. Thus E is associated with J_0 and H with J_1, when **H** is circular; conversely when **E** is circular.

We have first Fourier's cylinder function

$$\mathrm{J}_{0r}=\mathrm{J}_0(sr)=1-\frac{(sr)^2}{2^2}+\frac{(sr)^4}{2^2 4^2}-\ldots; \quad . \quad (305)$$

and its companion, which call G_0, is

$$\left.\begin{aligned} &\mathrm{G}_{0r}=\mathrm{G}_0(sr)=(2/\pi)[\mathrm{J}_{0r}\log sr+\mathrm{L}_{0r}], \\ &\text{where} \\ &\mathrm{L}_{0r}=\frac{(sr)^2}{2^2}-(1+\tfrac{1}{2})\frac{(sr)^4}{2^2 4^2}+(1+\tfrac{1}{2}+\tfrac{1}{3})\frac{(sr)^6}{2^2 4^2 6^2}-\ldots \end{aligned}\right\} (306)$$

The coefficient $2/\pi$ is introduced to simplify the solutions. The function $\mathrm{J}_1(sr)$ or J_{1r} is the negative of the first derivative of J_{0r} with respect to sr. Let $\mathrm{G}_1(sr)$ or G_{1r} be the function similarly derived from G_{0r}. The conjugate property, to be repeatedly used, is

$$(\mathrm{J}_0\mathrm{G}_1-\mathrm{J}_1\mathrm{G}_0)_r=-2/\pi sr. \quad . \quad . \quad . \quad (307)$$

We have also Stokes's formula for J_{0r}, useful when sr is real and not too small, viz.

$$\mathrm{J}_{0r}=(\pi sr)^{-\frac{1}{2}}[\mathrm{R}\,(\cos+\sin)\,sr+\mathrm{S}i\,(\sin-\cos)\,sr], \quad . \quad (308)$$

where R and Si are functions of sr to be presently given. The corresponding formula for G_{0r} is obtained by changing cos to sin and sin to $-$cos in (308).

Besides these two sets of solutions, we sometimes require to use a third set. A pair of solutions of the J_0 equation is

$$\left.\begin{aligned} &U=r^{-\frac{1}{2}}\epsilon^{qr}(R+S),\quad W=r^{-\frac{1}{2}}\epsilon^{-qr}(R-S),\\ \text{where}\quad &R\pm S=1\pm\frac{1}{8qr}+\frac{1^2 3^2}{\underline{|2}(8qr)^2}\pm\frac{1^2 3^2 5^2}{\underline{|3}(8qr)^3}+\dots \end{aligned}\right\}\quad(309)$$

The last also defines the R and Si in (308). R is real whether q^2 be + or −, whilst S is unreal when q^2 is −, or Si is then real, s^2 being +.

When qr is a + numeric, the solution U is meaningless, as its value is infinity. But in our investigations q^2 is a differential operator, so that the objection to U on that score is groundless. We shall use it to calculate the shape of an inward progressing wave, whilst W goes to find an outward wave. The results are fully convergent within certain limits of r and t. From this alone we see that a comprehensive theory of ordinary linear differential equations is sometimes impossible. They must be generalized into partial differential equations before they can be understood.

The conjugate property of U and W is

$$UW'-U'W=-2q/r, \quad . \quad . \quad . \quad . \quad (310)$$

if the $'=d/dr$. An important transformation sometimes required is

$$J_{0r}-iG_{0r}=2iW(2\pi q)^{-\frac{1}{2}}; \quad . \quad . \quad . \quad . \quad (311)$$

or, which means the same,

$$W=-\left(\frac{2q}{\pi}\right)^{\frac{1}{2}}[J_{0r}\log qr+L_{0r}]. \quad . \quad . \quad (312)$$

When we have obtained the differential equation in any problem, the assumption $s^2=$ a + constant converts it into the solution due to impressed force sinusoidal with respect to t and z; this requires $d^2/dz^2=-m^2$, and $d^2/dt^2=-n^2$, where m and n are positive constants, being 2π times the wave-shortness along z and 2π times the frequency of vibration respectively.

After (309) we became less exclusively mathematical. To go further in this direction, and come to electromagnetic waves, observe that we need not concern ourselves at all with F the radial component, in seeking for the proper differential equation connected with a surface of curl of impressed force; it is E and H only that we need consider, as the boundary conditions concern them. The second of (302) derives F from H.

When **H** is circular, the operator E/H is given by

$$\frac{\mathrm{E}}{\mathrm{H}}=\frac{s}{4\pi k+cp}\frac{\mathrm{J}_{0r}-y\mathrm{G}_{0r}}{\mathrm{J}_{1r}-y\mathrm{G}_{1r}}, \quad . \quad . \quad . \quad . \quad (313)$$

where y is undetermined. When E is circular, the operator E/H is given by

$$\frac{\mathrm{E}}{\mathrm{H}}=\frac{s}{4\pi k+cp}\frac{\mathrm{J}_{1r}-y\mathrm{G}_{1r}}{\mathrm{J}_{0r}-y\mathrm{G}_{0r}}. \quad . \quad . \quad . \quad . \quad (314)$$

The use of these operators greatly facilitates and systematizes investigation. The meaning is that (313) or (314) is the characteristic equation connecting E and H.

51. *Longitudinal Impressed E.M.F. in a thin Conducting Tube.*—Let an infinitely long thin conducting tube of radius a have conductance K per unit of its surface to longitudinal current, and be bounded by a dielectric on both sides. Strictly speaking the tube should be infinitely thin, in order to obtain instantaneous magnetic penetration, and yet be of finite conductance without possessing infinite conductivity, because that would produce opacity. In this tube let impressed electric force, of intensity e per unit length, act longitudinally, e being any function of t and z. We have to connect e with E and H internally and externally.

The magnetic force being circular, (313) is the resistance operator required. Within the tube take $y=0$ if the axis is to be included; else find y by some internal boundary condition. Outside the tube take $y=i$ when the medium is homogeneous and boundless, because that is the only way to prevent waves from coming from infinity; else find y by some outer boundary condition. There is no difficulty in forming the y to suit any number of coaxial cylinders possessing different electrical constants, by the continuity of E and H at each boundary, which equalizes the E/H's of its two sides, and so expresses the y of one side in terms of that on the other; but this is useless for our purpose. For the present take $y=0$ inside, and leave it unstated outside.

At $r=a$, E_a has the same value on both sides of the tube, on account of its thinness. In the substance of the tube $e+\mathrm{E}_a$ is the force of the flux. On the other hand H is discontinuous at the tube, thus

$$4\pi\mathrm{K}(e+\mathrm{E}_a)=\mathrm{H}_{(\mathrm{out})}-\mathrm{H}_{(\mathrm{in})}=\left(\frac{\mathrm{H}}{\mathrm{E}}(\mathrm{out})-\frac{\mathrm{H}}{\mathrm{E}}(\mathrm{in})\right)\mathrm{E}_a \quad . \quad (315)$$

In this use (313), and the conjugate property (307), and we at once obtain

$$e=\left[-1+\frac{4\pi k+cp}{4\pi\mathrm{K}s}\frac{2y}{\pi sa}\frac{1}{\mathrm{J}_{0a}(\mathrm{J}_{0a}-y\mathrm{G}_{0a})}\right]\mathrm{E}_a, \quad . \quad (316)$$

from which all the rest follows. Merely remarking concerning k that the realization of (316) when k is finite requires the splitting up of the Bessel functions into real and imaginary parts, that the results are complex, and that there are no striking peculiarities readily deducible; let us take $k=0$ at once, and keep to non-conducting dielectrics. Then, from (316), follow the equations of E and H, in and out; thus

$$E_{(in)} \text{ or }_{(out)} = \frac{J_{0r}(J_{0a}-yG_{0a}) \text{ or } J_{0a}(J_{0r}-yG_{0r})}{\frac{cp}{4\pi Ks}\frac{2y}{\pi sa} - J_{0a}(J_{0a}-yG_{0a})} e, \quad . \quad . \quad (317)$$

$$H_{(in)} \text{ or }_{(out)} = \frac{cp}{s} \frac{J\ \ (J_{0a}-yG_{0a}) \text{ or } J_{0a}(J_{1r}-yG_{1r})}{\text{same denominator}}, \quad . \quad . \quad (318)$$

which we can now examine in detail.

52. *Vanishing of External Field.* $J_{0a}=0$.—The very first thing to be observed is that $J_{0a}=0$ makes E and H and therefore also F vanish outside the tube, and that this property is independent of y, or of the nature of the external medium. We require the impressed force to be sinusoidal or simply periodic with respect to z and t, thus

$$e = e_0 \sin(mz+\alpha) \sin(nt+\beta), \quad . \quad . \quad . \quad (319)$$

so that ultimately

$$s^2 = n^2/v^2 - m^2; \quad . \quad . \quad . \quad . \quad . \quad . \quad . \quad (320)$$

and any one of the values of s given by $J_{0a}=0$ causes the evanescence of the external field. The solutions just given reduce to

$$\left.\begin{aligned} H &= -4\pi K(J_{1r}/J_{1a})e \\ E &= (s/cn)4\pi K(J_{0r}/J_{1a})ie \\ F &= -(cn)^{-1}4\pi K(J_{1r}/J_{1a})i(de/dz) \end{aligned}\right\} \quad . \quad . \quad . \quad (321)$$

which are fully realized, because i signifies p/n, or involves merely a time-differentiation performed on the e of (319).

The electrification is solely upon the inner surface of the tube. In its substance H falls from $-4\pi Ke$ inside to zero outside, and E_a being zero, the current in the tube is Ke per unit surface.

The independence of y raises suspicion at first that (321) may not represent the state which is tended to after e is started. But since the resistance of the tube itself is sufficient to cause initial irregularities to subside to zero, even were there a perfectly reflecting barrier outside the tube to prevent dissipation of these irregularities in space, there seems no reason to doubt that (321) do represent the state asymptotically tended to. Changing the form of y will only change the manner of the settling down. We may commence to change the nature of the medium immediately at the outer boundary of the tube. We cannot, however, have those abrupt assumptions of the steady or simply periodic state which characterize spherical waves, owing to the geometrical conditions of a cylinder.

53. *Case of two Coaxial Tubes.*—If there be a conducting tube anywhere outside the first tube, there is no current in it, except initially. From this we may conclude that if we transfer the impressed force to the outer tube, there will be no current in the inner. Thus, let there be an outer tube at $r=x$, of conductance K_1 per unit area, containing the impressed force e_1. We have

$$E_x=\frac{4\pi K_1 e_1}{Y_3-Y_2-4\pi K_1}, \quad . \quad . \quad . \quad . \quad . \quad . \quad (322)$$

where Y_3 and Y_2 are the H/E operators just outside and inside the tube, whilst E_x is the E at x, on either side of the tube, resulting from e_1. We have

$$Y_3=\frac{cp}{s}\frac{J_{1x}-y_1G_{1x}}{J_{0x}-y_1G_{0x}}, \quad Y_2=\frac{cp}{s}\frac{J_{1x}-yG_{1x}}{J_{0x}-yG_{0x}}, \quad . \quad . \quad (323)$$

where y_1 is settled by some external and y by some internal condition. In the present case the inner tube at $r=a$, if it contains no impressed force, produces the condition

$$Y_2-Y_1=4\pi K \text{ at } r=a, \quad . \quad . \quad . \quad . \quad (324)$$

where Y_1 is the internal H/E operator. Or

$$4\pi K=\frac{cp}{s}\left(\frac{J_{1a}-yG_{1a}}{J_{0a}-yG_{0a}}-\frac{J_{1a}}{J_{0a}}\right),$$

giving

$$y=\frac{4\pi KJ^2_{0a}}{\frac{2}{\pi sa}\frac{cp}{s}+4\pi KJ_{0a}G_{0a}}. \quad . \quad . \quad . \quad . \quad . \quad (325)$$

Now, using (323) in (322) brings it to

$$E_x=\frac{(J_{0x}-yG_{0x})(J_{0x}-y_1G_{0x})4\pi K_1 e_1}{\frac{cp}{s}(y_1-y)\frac{2}{\pi sx}-4\pi K_1(J_{0x}-yG_{0x})(J_{0x}-y_1G_{0x})}, \quad (326)$$

in which y is given by (325), and from (326) the whole state due to e_1 follows, as modified by the inner tube.

Now $J_{0a}=0$ makes $y=0$; this reduces (326) to

$$E_x = \frac{J_{0x}(J_{0x} - y_1 G_{0x}) 4\pi K_1 e_1}{\frac{cp}{s} y_1 \frac{2}{\pi s x} - 4\pi K_1 J_{0x}(J_{0x} - y_1 G_{0x})}; \quad . \quad . \quad . \quad (327)$$

and, by comparison with (317) we see that it is now the same as if the inner tube were non-existent. That is, when it is situated at a nodal surface of E due to impressed force in the outer tube, and there is therefore no current in it (except transversely, to which the dissipation of energy is infinitely small), its presence does nothing, or it is perfectly transparent.

It is clearly unnecessary that the external impressed force should be in a tube. Let it only be in tubular layers, without specification of actual distribution or of the nature of the medium, except that it is in layers so that c, k, and μ are functions of r only; then if the axial portion be nonconducting dielectric, the J_{0r} function specifies E and allows there to be nodal surfaces, for instance $J_{0a}=0$, where a conducting tube may be placed without disturbing the field. Admitting this property *ab initio*, we can conversely conclude that e in the tube at $r=a$ will, when $J_{0a}=0$, make *every* external cylindrical surface a nodal surface, and therefore produce no external disturbance at all.

54. Now go back to § 51, equations (317) (318). There are no *external* nodal surfaces of E in general (exception later). We cannot therefore find a place to put a tube so as not to disturb the existing field due to e in the tube at $r=a$. But we may now make use of a more general property. To illustrate simply, consider first the electromagnetic theory of induction between linear circuits. Let there be any number of circuits, all containing impressed forces, producing a determinate varying electromagnetic field. In this field put an additional circuit of infinite resistance. The E.M.F. in it, due to the other circuits, will cause no current in it of course, so that no change in the field takes place. Now, lastly, close the circuit or make its resistance finite, and simultaneously put in it impressed force which is at every moment the negative of the E.M.F. due to the other circuits. Since no current is produced there will still be no change, or everything will go on as if the additional circuit were non-existent.

Applying this to our tubes, we may easily verify by the previous equations that when there are two coaxial tubes, both containing impressed forces, we can reduce the resultant electromagnetic field everywhere to that due to the impressed force in one tube, provided we suitably choose the impressed force in the second to be the negative of the electric force of field due to e in the first tube when the second is non-existent. That is, we virtually abolish the conductance of the second tube and make it perfectly transparent.

55. *Perfectly Reflecting Barrier. Its effects. Vanishing of Conduction Current.*—To produce nodal surfaces of E outside the tube containing the vibrating impressed force, we require an external barrier, which shall prevent the passage of energy or its absorption, by wholly reflecting all disturbances which reach it. Thus, let there be a perfect conductor at $r=x$. This makes $E=0$ there. This requires that the y in (317), (318) shall have the value J_{0x}/G_{0x}, whereas without any bound to the dielectric it would be i. We can now choose m and n so as to make $J_{0x}=0$. This reduces those equations to

$$\text{(in and out)} \quad \left.\begin{aligned} E &= -\frac{J_{0r}}{J_{0a}}e; \quad F = -\frac{1}{s}\frac{J_{1r}}{J_{0a}}e; \\ H &= -\frac{1}{s}\frac{J_{1r}}{J_{0a}}cpe. \end{aligned}\right\} \quad . \quad . \quad . \quad (328)$$

This solution is now the same inside and outside the tube containing the impressed force, and there is no current in the tube, that is, no longitudinal current.

To understand this case, take away the impressed force and the tube. Then (328) represents a conservative system in stationary vibration. Now, by the preceding, we may introduce the tube at a nodal surface of E without disturbing matters, provided there be no impressed force in the tube. But if we introduce the tube anywhere else, where E is not zero, we require, by the preceding, an impressed force which is at every moment the negative of the undisturbed force of the field, in order that no change shall occur. Now this is precisely what the solution (328) represents, e in the tube being cancelled by the force of the field, so that there is no conduction-current. The remarkable thing is that it is the impressed force in the tube itself that sets up the vibrating field, and gradually ceases to work, so that in the end it and the tube may be removed without altering the field. That a perfect conductor as reflector is required is a detail of no moment in its theoretical aspect.

Shifting the tube, with a finite impressed force in it, towards a nodal suface of E, sends up the amplitude of the vibrations to any extent.

56. $K=0$ *and* $K=\infty$.—If the tube have no conductance, e produces no effect. This is because the two surfaces of

curl of e are infinitely close together, and therefore cancel, not having any conductance between them to produce a discontinuity in the magnetic force.

But if the tube have infinite conductance, we produce complete independence between the internal and external fields, except in the quite unessential particular that the two surfaces of curl e are of opposite kind and time together. Equations (317), (318) reduce to

$$\text{(in)} \quad \mathrm{E} = -\frac{\mathrm{J}_{0r}}{\mathrm{J}_{1a}} e, \quad \mathrm{F} = -\frac{1}{s}\frac{\mathrm{J}_{1r}}{\mathrm{J}_{0a}} e, \quad \mathrm{H} = -\frac{1}{s}\frac{\mathrm{J}_{1r}}{\mathrm{J}_{0a}} cpe. \quad . \quad (329)$$

$$\text{(out)} \left\{ \begin{array}{l} \mathrm{E} = -\dfrac{\mathrm{J}_{0r} - y\mathrm{G}_{0r}}{\mathrm{J}_{0a} - y\mathrm{G}_{0a}} e, \quad \mathrm{F} = \dfrac{1}{s}\dfrac{\mathrm{J}_{1r} - y\mathrm{G}_{1r}}{\mathrm{J}_{0a} - y\mathrm{G}_{0a}}, \\ \mathrm{H} = -\dfrac{1}{s}\dfrac{\mathrm{J}_{1r} - y\mathrm{G}_{1r}}{\mathrm{J}_{0a} - y\mathrm{G}_{0a}} cpe. \end{array} \right\} \quad . \quad (330)$$

Observe that (329) is the same as (328). The external solution (330) requires y to be stated. When $y = i$, for a boundless dielectric, the realization is immediate.

57. $s=0$. *Vanishing of* E *all over, and of* F *and* H *also internally.*—This is a singularity of quite a different kind. When $n = mv$, we make $s = 0$. Of course there is just one solution with a given wave-length along z: a great frequency with small wave-length, and conversely.

E vanishes all over, that is both inside and outside the tube containing e, provided s/y is zero. The internal H and therefore also F vanish. Thus within the tube is no disturbance, and outside, (317) (318) reduce to

$$\text{(out)} \quad \mathrm{H} = \frac{a}{r} 4\pi \mathrm{K} e, \quad \mathrm{F} = \frac{1}{cn}\frac{a}{r} 4\pi \mathrm{K} i \frac{de}{dz}. \quad . \quad . \quad . \quad . \quad (331)$$

Observe that H and F do not fluctuate or alternate along r, but that H has the same distribution (out from the tube) as if e were steady and did not vary along z.

A special case is $m=0$. Then also $n=0$, or e is steady and independent of z. F vanishes, and the first of (331) expresses the steady state.

Without this restriction, the current in the tube is Ke per unit surface, owing to the vanishing of the opposing longitudinal E of the field. This property was, by inadvertence, attributed by me in a former paper * to a wire instead of a

* "On Resistance and Conductance Operators," Phil. Mag. Dec. 1887, p. 492, Ex. *j*.

tube. The wave-length must be great in order to render it applicable to a wire, because instantaneous penetration is assumed.

I mentioned that s/y must vanish. This occurs when $y=i$, or the external dielectric is boundless. But it also occurs when $E=0$ at $r=x$, produced by a perfectly conductive screen. This is plainly allowable because it does not interfere with the $E=0$ all over property. What the screen does is simply to terminate the field abruptly. Of course it is electrified.

58. $s=0$ *and* $H_x=0$.—But with other boundary conditions, we do not have the solutions (331). Thus, let $H_x=0$, instead of $E_x=0$. This makes $y=J_{1x}/G_{1x}$ in (317), (318). There are at least two ways (theoretical) of producing this boundary condition. First, there may be at $r=x$ a screen made of a perfect magnetic conductor ($g=\infty$). Or, secondly, the whole medium beyond $r=x$ may be infinitely elastive and resistive ($c=0$, $k=0$) to an infinite distance.

Now choose $s=0$ in addition and reduce (317), (318). The results are

$$\left.\begin{aligned} E&=-\frac{e}{1+\frac{1}{2}x^2cp/4\pi Ka}; \quad F=-\frac{1}{cp}\frac{dH}{dz}, \\ \text{(in) or (out) } H&=-\frac{cpe}{1+\frac{1}{2}x^2cp/4\pi Ka}\left(\frac{r}{2}\text{ or }\frac{r}{2}-\frac{x^2}{2r}\right), \end{aligned}\right\} \quad (332)$$

which are at once realized by removing p from the denominator to the numerator.

Although E is not now zero, it is independent of r, only varying with t and z.

When s^2 is negative, or $n<m/v$, the solutions (317), (318) require transforming in part because some of the Bessel functions are unreal. Use (312), because q is now real. There are no alternations in E or H along r. They only commence when $n>mv$.

59. *Separate actions of the two surfaces of curl* **e**.—Since all the fluxes depend solely upon the curl of **e** and not upon its distribution, and there are two surfaces of curl **e** in the tube problem, their actions, which are independent, may be separately calculated. The inner surface may arise from **e** in the − direction in the inner dielectric, or by the same in the + direction in the tube and beyond it. The outer may be due to **e** in the − direction beyond the tube, or in the + direction in the tube and inner dielectric.

We shall easily find that the inner surface of curl of **e**, say of surface density f_1, produces

$$\left.\begin{array}{ll}\text{(in)} & E=J_{0r}\dfrac{(J_{1a}-yG_{1a})-(J_{0a}-yG_{0a})4\pi Ks/cp}{2y/\pi sa-J_{0a}(J_{0a}-yG_{0a})4\pi Ks/cp}f_1 \\ \text{(out)} & E=\dfrac{J_{1a}(J_{0r}-yG_{0r})}{\text{same denominator}}f_1\end{array}\right\} . \quad (333)$$

from which H may be got by the E/H operator.

The external sheet, say f_2, produces

$$\left.\begin{array}{ll}\text{(in)} & E=\dfrac{J_{0r}(J_{1a}-yG_{1a})}{\cdot\;\cdot\;\cdot\;\cdot\;\cdot}f_2, \\ \text{(out)} & E=(J_{0r}-yG_{0r})\,\dfrac{J_{1a}+J_{0a}4\pi Ks/cp}{\cdot\;\cdot\;\cdot\;\cdot\;\cdot\;\cdot}f_2,\end{array}\right\} . \quad (334)$$

where the unwritten denominators are as in the first of (333). Observe that when $J_{1a}=0$, f_1 produces no external field (in tube or beyond it). It is then only f_2 that operates in the tube and beyond.

Now take $f_2=e$ and $f_1=-e$ in (333) and (334) and add the results. We then obtain (317), (318); and it is now $J_{0a}=0$ that makes the external field vanish, instead of $J_{1a}=0$ when f_1 alone is operative.

Having treated this problem of a tube in some detail, the other examples may be very briefly considered, although they too admit of numerous singularities.

60. *Circular Impressed Force in Conducting-tube.*—The tube being as before, let the impressed force **e** (per unit length) act circularly in it instead of longitudinally, and let **e** be a function of t only, so that we have an inner and an outer cylindrical surface of longitudinally directed curl of **e**. **H** is evidently longitudinal and **E** circular, so that we now require to use the (314) operator.

At the tube E_a is continuous, this being the tensor of the force of the flux on either side, and H is discontinuous thus,

$$H_{(in)}-H_{(out)}=4\pi K(e+E_a),$$

or

$$e=-\left\{1+\frac{1}{4\pi K}\left(\frac{H}{E}\,(\text{out})-\frac{H}{E}\,(\text{in})\right)\right\}E_a. \quad . \quad (335)$$

Substituting the (314) operator, with $y=0$ inside, and y undetermined outside, and using the conjugate property (307), we obtain

$$H_{(in)}\text{ or }_{(out)}=-i\,\frac{(J_{1a}-yG_{1a})J_{0r}\text{ or }J_{1a}(J_{0r}-yG_{0r})}{\mu vJ_{1a}(J_{1a}-yG_{1a})+\dfrac{y}{4\pi K}\,\dfrac{2v}{\pi ap}}\,e, \quad (336)$$

$$E_{(in)}\text{ or }_{(out)}=-\mu v\,\frac{(J_{1a}-yG_{1a})J_{1r}\text{ or }J_{1a}(J_{1r}-yG_{1r})}{\text{same denominator}}\,e. \quad (337)$$

When e is simply periodic, $J_{1a}=0$ makes the external E and H vanish independent of the nature of y. The complete solution is then

$$H_{(in)}=4\pi K\frac{J_{0r}}{J_{0a}}e, \quad E_{(in)}=-4\pi K\mu v\frac{J_{1r}}{J_{0a}}ie. \quad . \quad (338)$$

The conduction-current in the tube is Ke per unit area of surface.

To make the conduction-current vanish by balancing the impressed force against the electric force of the field that it sets up, put an infinitely conducting screen at $r=x$ outside the tube and choose the frequency to make $J_{1x}=0$, since we now have $y=J_{1x}/G_{1x}$. We shall then have the same solution inside and outside, viz.

$$H=-\frac{1}{\mu v}\frac{J_{0r}}{J_{1a}}ie, \quad E=-\frac{J_{1r}}{J_{1a}}e; \quad . \quad . \quad . \quad (339)$$

so that at the tube itself $E=-e$. This case may be interpreted as in § 55, the tube being at a nodal surface of E.

A special case of (338) is when $n=0$, or e is steady. Then there is merely the longitudinal **H** inside the tube, given by

$$H=4\pi Ke.$$

61. *Cylinder of longitudinal curl of* **e** *in a Dielectric.*—In a nonconductive dielectric let the impressed electric force be such that its curl is confined to a cylinder of radius a, in which it is uniformly distributed, and is longitudinal. Let f be the tensor of curl **e**, and let it be a function of t only. Since **E** is circular and **H** longitudinal, we have (314) as operator, in which k is to be zero. This is outside the cylinder. Inside, on the other hand, on account of the existence of curl **e**, the equation corresponding to (314) is

$$\frac{E}{H-f/\mu p}=\frac{s}{cp}\frac{J_{1r}}{J_{0r}}. \quad . \quad . \quad . \quad . \quad . \quad (340)$$

At the boundary $r=a$ both E and H are continuous; so, by taking $r=a$ in (340) and in the corresponding (314) with $k=0$, and eliminating E_a or H_a between them, we obtain the equation of the other. We obtain

$$(\text{out})\left\{\begin{array}{l} E=\frac{1}{2}\pi ay^{-1}J_{1a}(J_{1r}-yG_{1r})f, \\ H=\frac{1}{2}\pi ay^{-1}J_{1a}(J_{0r}-yG_{0r})(\mu v)^{-1}if, \end{array}\right\} \quad . \quad . \quad (341)$$

in which y, as usual, is to be fixed by an external boundary condition, or, if the medium be boundless, $y=i$.

We see at once that $J_{1a}=0$, with f simply periodic, makes the external fluxes vanish. We should not now say that it makes the external field vanish, though the statement is true as regards H, because the electric force of the field does not vanish; it cancels the impressed force, so that there is no flux. This property is apparently independent of y. But, since there is no resistance concerned, except such as may be expressed in y, it is clear that (341) sinusoidally realized cannot represent the state which is tended to after starting f, unless there be either no barrier, so that initial disturbances can escape, or else there be resistance somewhere, to be embodied in y, so that they can be absorbed, though only through an infinite series of passages between the boundary and the axis of the initial wave and its consequences.

Thus, with a conservative barrier producing $E=0$ at $r=x$, and $y=J_{1x}/G_{1x}$, there is no escape for the initial effects, which remain in the form of free vibrations, whilst only the forced vibrations are got by taking $s^2=+$ constant in (341). The other part of the solution must be separately calculated. If $J_{1x}=0$, E and H run up infinitely. If $J_{1a}=0$ also, the result is ambiguous.

With no barrier at all, or $y=i$, we have

$$\text{out}\left\{\begin{array}{l} E=-(2a)^{-1}J_{1a}(G_{1r}+iJ_{1r})f_0, \\ H=(2a\mu v)^{-1}J_{1a}(J_{0r}-iG_{0r})f_0, \end{array}\right\} \quad . \quad . \quad (342)$$

which are fully realized. Here $f_0=f\pi a^2$, which may be called the strength of the filament. We may most simply take the impressed force to be circular, its intensity varying as r within and inversely as r outside the cylinder. Then $f=2e\,/a$, if e_a is the intensity at $r=a$.

When nr/v is large, (342) becomes, by (308),

$$(\text{out})\ E=\mu v H=\frac{f_0 n}{4v}\left(\frac{2v}{\pi nr}\right)^{\frac{1}{2}}\sin\left(nt-\frac{nr}{v}+\frac{\pi}{4}\right) \quad . \quad (343)$$

approximately. $2\pi r$ should be a large multiple, and $2\pi a$ a small fraction of the wave-length along r.

62. *Filament of curl* **e**. *Calculation of Wave.*—In the last let f_0 be constant whilst a is made infinitely small. It is then a mere filament of curl of **e** at the axis that is in operation. We now have, by the second of (342), with $J_{1a}=\frac{1}{2}na/v$,

$$H=-(cp/4)(iJ_{0r}+G_{0r})f_0, \quad . \quad . \quad . \quad . \quad (344)$$

which may be regarded as the simply periodic solution or as the differential equation of H. In the latter case, put in terms of W by (311), then

$$H=(2\mu v)^{-1}(q/2\pi)^{\frac{1}{2}}Wf_0; \quad . \quad . \quad . \quad . \quad (345)$$

or, expanding by (309),

$$\mathrm{H}=\frac{1}{2\mu v}\frac{1}{(2\pi r)^{\frac{1}{2}}}\epsilon^{-qr}\left(1-\frac{1}{8qr}+\frac{1^2 3^2}{\lfloor 2(8qr)^2}-\dots\right)q^{\frac{1}{2}}f_0, \quad (346)$$

in which f_0 may be any function of the time. Let it be zero before and constant after $t=0$. Then, first,

$$q^{\frac{1}{2}}f_0=f_0(\pi vt)^{-\frac{1}{2}}. \quad \dots \dots \quad (347)$$

Next effect the integrations of this function indicated by the inverse powers of q or p/v, thus

$$f_0\left(1-\frac{1}{8qr}+\dots\right)(\pi vt)^{-\frac{1}{2}}=\left(1-\tfrac{1}{2}\left(\frac{vt}{2r}\right)+\frac{1\,.\,3}{2^2\lfloor 2}\left(\frac{vt}{2r}\right)^2-\dots\right)(\pi vt)^{-\frac{1}{2}}$$

$$=(1+vt/2r)^{-\frac{1}{2}}(\pi vt)^{-\frac{1}{2}}=(2r/\pi)^{\frac{1}{2}}[vt(vt+2r)]^{-\frac{1}{2}}. \quad (348)$$

Lastly, operating on this by ϵ^{-qr} turns vt to $vt-r$, and brings (346) to

$$\mathrm{H}=(f_0/2\pi\mu v)(v^2t^2-r^2)^{-\frac{1}{2}}, \quad \dots \quad (349)$$

which is ridiculously simple. Let Z be the time-integral of H, then

$$\mathrm{Z}=\frac{cf_0}{2\pi}\log\left[\frac{vt}{r}+\left(\frac{v^2t^2}{r^2}-1\right)^{\frac{1}{2}}\right], \quad \dots \quad (350)$$

from which we may derive E ; thus

$$\text{curl }\mathbf{Z}=c\mathbf{E}, \quad \text{or } \mathrm{E}=-\frac{1}{c}\frac{d\mathrm{Z}}{dr}=\frac{vtf_0}{2\pi r(v^2t^2-r^2)^{\frac{1}{2}}}. \quad (351)$$

The other vector-potential **A**, such that $\mathbf{E}=-p\mathbf{A}$ is obviously

$$\mathrm{A}=-\frac{1}{2\pi v}\left(\frac{v^2t^2}{r^2}-1\right)^{\frac{1}{2}}f_0 \quad \dots \quad (352)$$

All these formulæ of course only commence when vt reaches r. The infinite values of E and H at the wave-front arise from the infinite concentration of the curl of e at the axis.

Notice that

$$\mathrm{E}=\mathrm{H}t/rc \quad \dots \dots \quad (353)$$

everywhere. It follows from this connexion between E and H (or from their full expressions) that

$$c\mathrm{E}^2-\mu\mathrm{H}^2=ce^2=c(f_0/2\pi r)^2; \quad \dots \quad (354)$$

where e denotes the intensity of impressed force at distance r, when it is of the simplest type, above described. That is, the excess of the electric over the magnetic energy at any point is independent of the time. Both decrease at an equal rate; the magnetic energy to zero, the electric energy to that of the final steady displacement $ce/4\pi$.

The above E and H solutions are fundamental, because all electromagnetic disturbances due to impressed force depend solely upon, and come from, the lines of curl of the impressed force. From them, by integration, we can find the disturbances due to any collection of rectilinear filaments of **f**. Thus, to find the H due to a plane sheet of parallel uniformly distributed filaments, of surface-density f, we have, by (349), at distance a from the plane, on either side,

$$\mathrm{H}=\int \frac{f dy}{2\pi\mu v(v^2t^2-a^2-y^2)^{\frac{1}{2}}}=\frac{f}{2\pi\mu v}\left[\sin^{-1}\frac{y}{(v^2t^2-a^2)^{\frac{1}{2}}}\right],$$

where the limits are $\pm(v^2t^2-a^2)^{\frac{1}{2}}$. Therefore

$$\mathrm{H}=f/2\mu v$$

after the time $t=a/v$; before then, H is zero.

Similarly, a cylindrical sheet of longitudinal **f** produces

$$\mathrm{H}=\frac{fa}{2\pi\mu v}\int \frac{d\theta}{(v^2t^2-b^2)^{\frac{1}{2}}};$$

where b is the distance of the point where H is reckoned from the element $a\,d\theta$ of the circular section of the sheet, a being its radius. The limits have to be so chosen as to include all elements of f which have had time to produce any effect at the point in question. When the point is external and vt exceeds $a+r$ the limits are complete, viz. to include the whole circle. The result is then, at distance r from the axis of the cylinder,

$$\mathrm{H}=\frac{fa/\mu v}{(v^2t^2-a^2-r^2)^{\frac{1}{2}}}\left[1+\frac{1.3}{2^2\underline{|2}}\frac{x}{2}+\frac{1.3.5.7}{2^4\underline{|4}}\frac{x^2}{2^3}\frac{4.3}{1.2}\right.$$
$$\left.+\frac{1.3.5.7.9.11}{2^6\underline{|6}}\frac{x^3}{2^5}\frac{6.5.4}{1.2.3}+\dots\right], \quad (355)$$

where

$$x=(2ar)^{\frac{1}{2}}(v^2t^2-a^2-r^2)^{-\frac{1}{2}}.$$

This formula begins to operate when $x=1$, or $vt=a+r$. As time goes on, x falls to zero, leaving only the first term.

[To be continued.]

From the PHILOSOPHICAL MAGAZINE for December 1888.

On Electromagnetic Waves, especially in relation to the Vorticity of the Impressed Forces; and the Forced Vibrations of Electromagnetic Systems. By OLIVER HEAVISIDE.

[Concluded from p. 449.]

63. *CYLINDRICAL Surface of circular curl* **e** *in a Dielectric.*—Let the curl of the impressed electric force be wholly situated on the surface of a cylinder of radius a in a nonconducting dielectric. The impressed force **e** to correspond may then be most conveniently imagined to be either longitudinal, within or without the cylinder, uniformly distributed in either case (though oppositely directed), and the density of curl **e** will be e; or, the impressed force may be transferred to the surface of the cylinder, by making **e** radial, but confined to an infinitely thin layer. The measure of the surface-density of curl **e** will now be

$$f=\frac{de}{dz},=\mathrm{E}_{(\mathrm{out})}-\mathrm{E}_{(\mathrm{in})}, \quad . \quad . \quad . \quad . \quad (356)$$

where e is the total impressed force (its line-integral through the layer). The second form of this equation shows the effect produced on the electric force **E** of the flux, outside and inside the surface. This **E** is, as it happens, also the force of the field ; but in the other case, when **e** is uniformly distributed within the cylinder, producing $f=e$, we have the same discontinuity produced by f.

H being circular, we use the operator (313). Applying it to (356) we obtain

$$f=\frac{s}{cp}\left(\frac{J_{0a}}{J_{1a}}-\frac{J_{0a}-yG_{0a}}{J_{1a}-yG_{1a}}\right)H_a; \quad . \quad . \quad . \quad (357)$$

from which, by the conjugate property (307), and the operator (313), we derive

$$E_{(in)} \text{ or }_{(out)}=\frac{\pi as}{2y}[J_{0r}(J_{1a}-yG_{1a}) \text{ or } J_{1a}(J_{0r}-yG_{0r})]f, \quad (358)$$

$$H_{(in)} \text{ or }_{(out)}=\frac{\pi acp}{2y}[J_{1r}(J_{1a}-yG_{1a}) \text{ or } J_{1a}(J_{1r}-yG_{1r})]f, \quad (359)$$

in which f is a function of t, and it may be also of z. If so, then we have the radial component F of electric force given by

$$F_{(in)} \text{ or }_{(out)}=-\frac{\pi a}{2y}[J_{1r}(J_{1a}-yG_{1a}) \text{ or } J_{1a}(J_{1r}-yG_{1r})]\frac{df}{dz}. \quad (360)$$

From these, by the use of Fourier's theorem, we can build up the complete solutions for any distribution of f with respect to z; for instance, the case of a single circular line of curl **e**.

64. $J_{1a}=0$. *Vanishing of external field.*—Let f be simply periodic with respect to t and z; then $J_{1a}=0$, or

$$J_1\{a\sqrt{n^2/v^2-m^2}\}=0, \quad . \quad . \quad . \quad . \quad (361)$$

produces evanescence of E and H outside the cylinder. The independence of this property of y really requires an unbounded external medium, or else boundary resistance, to let the initial effects escape or be dissipated, because no resistance appears in our equations except in y. The case $s=0$ or $n=mv$ is to be excepted from (361); it is treated later.

65. $y=i$. *Unbounded medium.*—When $n/v>m$, s is real, and our equations give at once the fully realized solutions in the case of no boundary, by taking $y=i$,

$$\left.\begin{aligned}
&H_{(in)} \text{ or }_{(out)}=\tfrac{1}{2}\pi acn[J_{1r}(J_{1a}-iG_{1a}) \text{ or } J_{1a}(J_{1r}-iG_{1r})]f,\\
&E_{(in)} \text{ or }_{(out)}=-\tfrac{1}{2}\pi as[J_{0r}(G_{1a}+iJ_{1a}) \text{ or } J_{1a}(G_{0r}+iJ_{0r})]f,\\
&F_{(in)} \text{ or }_{(out)}=\tfrac{1}{2}\pi a[J_{1r}(G_{1a}+iJ_{1a}) \text{ or } J_{1a}(G_{1r}+iJ_{1r})](df/dz),
\end{aligned}\right\} (362$$

in which i means p/n.

The instantaneous outward transfer of energy per unit length of cylinder is (by Poynting's formula)

$$-\frac{EH}{4\pi}\times 2\pi r,$$

and the mean value with respect to the time comes to

$$\frac{cn}{8\pi}(f_0\pi am\cos mz\,\mathrm{J}_{1a})^2, \quad . \quad . \quad . \quad . \quad (363)$$

if f_0 is the maximum value of f. This may of course be again averaged to get rid of the cosine.

66. $s=0$. *Vanishing of external* E.—When $n=mv$, we make $s=0$, and then (362) reduce to the singular solution

$$\left.\begin{aligned} &\mathrm{H}_{(\mathrm{in})}=\tfrac{1}{2}rcpf, && \mathrm{H}_{(\mathrm{out})}=\tfrac{1}{2}\frac{a^2}{r}cpf, \\ &\mathrm{E}_{(\mathrm{in})}=f, && \mathrm{E}_{(\mathrm{out})}=0, \\ &\mathrm{F}_{(\mathrm{in})}=-\tfrac{1}{2}r\frac{df}{dz}, && \mathrm{F}_{(\mathrm{out})}=-\tfrac{1}{2}\frac{a^2}{r}\frac{df}{dz}. \end{aligned}\right\} \quad . \quad (364)$$

Observe that the internal longitudinal displacement is produced entirely by the impressed force (*if* it be internal), though there is radial displacement also on account of the divergence of e (if internal). Outside the cylinder the displacement is entirely perpendicular to it.

H and F do not alternate along r. This is also true when s^2 is negative, or n lies between 0 and mv. Then, q^2 being positive, we have

$$\mathrm{E}_{(\mathrm{out})}=\tfrac{1}{2}a^2q^2\left(\frac{2}{sa}\mathrm{J}_{1a}\right)[\mathrm{J}_{0r}\log qr+\mathrm{L}_{0r}]f, \quad . \quad . \quad (365)$$

as the rational form of the equation of the external E when the frequency is too low to produce fluctuations along r.

The system (364) may be obtained directly from (358) to (360) on the assumption that s/y is zero when s is zero. But (364) appears to require an unbounded medium. Even in the case of the boundary condition $\mathrm{E}=0$ at $r=x$, which harmonizes with the vanishing of E externally in (364), there will be the undissipated initial effects continuing.

If, on the other hand, $\mathrm{H}_x=0$, making $y=\mathrm{J}_{0x}/\mathrm{G}_{0x}$, we shall not only have the undissipated initial effects, but a different form of solution for the forced vibrations. Thus, using this expression for y, and also $s=0$, in (358) to (360), we obtain

$$\left.\begin{aligned} &\mathrm{H}_{(\mathrm{in})}=\frac{a}{2}\left(1-\frac{a^2}{x^2}\right)\frac{r}{a}cpe\,; && \mathrm{H}_{(\mathrm{out})}=\frac{a}{2}\left(1-\frac{r^2}{x^2}\right)\frac{a}{r}cpe; \\ &\mathrm{E}_{(\mathrm{in})}=\left(1-\frac{a^2}{x^2}\right)e\,; && \mathrm{E}_{(\mathrm{out})}=-\frac{a^2}{x^2}e; \end{aligned}\right\} \quad . \quad (366)$$

representing the forced vibrations.

67. *Effect of suddenly starting a filament of* e.—The vibratory effects due to a vibrating filament we find by taking a

infinitely small in (362), that is $J_{1a}=\frac{1}{2}sa$. To find the wave produced by suddenly starting such a filament, transform equations (358), (359) by means of (311). We get

$$\left.\begin{aligned} E_{(in)} &= -(\pi/2q)^{\frac{1}{2}} a J_{0r} W_a' e, \\ E_{(out)} &= -\tfrac{1}{2}(\pi/2q^3)^{\frac{1}{2}} a^2 \left(\frac{2}{sa} J_{1a}\right) W e, \\ H_{(in)} &= -(\pi q/2)^{\frac{1}{2}} \frac{ar}{2\mu v}\left(\frac{2}{sr} J_{1r}\right) W_a' e, \\ H_{(out)} &= -(\pi q/2)^{\frac{1}{2}} \frac{a^2}{2\mu v}\left(\frac{2}{sa} J_{1a}\right) W' e; \end{aligned}\right\} \quad . \quad (367)$$

where W is given by (309); the accent means differentiation to r, and the suffix a means the value at $r=a$.

In these, let $e_0=\pi a^2 e$, which we may call the strength of the filament, and let a be infinitely small. We then obtain

$$(\text{out})\left\{\begin{aligned} H &= -(q/2\pi)^{\frac{1}{2}}(2\mu v)^{-1} W' e_0, \\ E &= \tfrac{1}{2}(q^3/2\pi)^{\frac{1}{2}} W e_0. \end{aligned}\right\} \quad . \quad . \quad (368)$$

Now if e_0 is a function of t only, it is clear that there is no scalar electric potential involved. We may therefore advantageously employ (and for a reason to be presently seen) the vector-potential **A**, such that

$$E=-pA, \quad \text{or } A=-p^{-1}E; \quad \text{and } \mu H=-\frac{dA}{dr}. \quad (369)$$

The equation of A is obviously, by the first of (369) applied to second of (368),

$$A=\tfrac{1}{2}(p/2\pi v^3)^{\frac{1}{2}} W e_0. \quad . \quad . \quad . \quad . \quad . \quad (370)$$

Comparing this equation with that of H in (345) (problem of a filament of curl of **e**), we see that f_0 there becomes e_0 here, and μH there becomes A here. The solution of (370) may therefore be got at once from the solution of (345), viz. (349). Thus

$$A=\frac{e_0}{2\pi v(v^2t^2-r^2)^{\frac{1}{2}}}; \quad . \quad . \quad . \quad . \quad . \quad (371)$$

from which, by (369),

$$E=\frac{e_0 vt}{2\pi(v^2t^2-r^2)^{\frac{3}{2}}}, \quad H=-\frac{e_0 r}{2\pi\mu v(v^2t^2-r^2)^{\frac{3}{2}}}, \quad . \quad (372)$$

the complete solution. It will be seen that

$$A=Et+r\mu H, \quad . \quad . \quad . \quad . \quad . \quad . \quad (373)$$

whilst the curious relation (353) in the problem of a filament of curl **e** is now replaced by

$$A = r\mu Z/t, \quad \ldots\ldots \quad (374)$$

where Z is the time-integral of the magnetic force; so that

$$H = pZ, \quad \text{and curl } \mathbf{Z} = c\mathbf{E}, \quad \ldots\ldots \quad (375)$$

Z being merely the vectorized Z. It is the vector-potential of the magnetic current.

The following reciprocal relation is easily seen by comparing the differential equations of an infinitely fine filament e_0 and a finite filament. The electric current-density at the axis due to a longitudinal cylinder of **e** (uniform) of radius a is numerically identical with the total current through the circle of radius a due to the same total impressed force (that is, $\pi a^2\mathbf{e}$) concentrated in a filament at the axis, at corresponding moments.

68. Having got the solutions (372) for a filament e_0, it might appear that we could employ them to build up the solutions in the case of, for instance, a cylinder of longitudinal impressed force of finite radius a. But, according to (372), E would be positive and H negative everywhere and at every moment, in the case of the cylinder, because the elementary parts are all positive or all negative. This is clearly a wrong result. For it is certain that, at the first moment of starting the longitudinal impressed force of intensity e in the cylinder, E just outside it is negative; thus

$$E = \pm \tfrac{1}{2} e, \text{ in or out, at } r = a,\ t = 0;$$

and that H is positive; viz.

$$H = e/2\mu v \text{ at } r = a,\ t = 0.$$

We know further that, as E starts negatively just outside the cylinder, E will be always negative at the front of the outward wave, and H positive; thus

$$-E = \mu v H = \tfrac{1}{2} e \times (a/r)^{\frac{1}{2}}, \quad \ldots\ldots \quad (376)$$

the variation in intensity inversely as the square root of the distance from the axis being necessitated in order to keep the energy constant at the wave-front. The same formula with $+E$ instead of $-E$ will express the state at the front of the wave running in to the axis. There is thus a momentary infinity of E at the axis, viz. when $t = a/v$.

So far we can certainly go. Less securely, we may conclude that during the recoil, E will be settling down to its steady value e within the cylinder, and therefore the force of the field there will be positive, and, by continuity, also positive outside the cylinder. Similarly, H must be negative at any distance within which E is decreasing. We conclude therefore that the filament solutions (372) only express the settling down to the final state, and are not comprehensive enough to be employed as fundamental solutions.

69. *Sudden starting of* **e** *longitudinal in a Cylinder.*—In order to fully clear up what is left doubtful in the last paragraph, I have investigated the case of a cylinder of **e** comprehensively. The following contains the leading points. We have to make four independent investigations: viz. to find (1) the initial inward wave; (2) the initial outward wave; (3) the inside solution after the recoil; (4) the outside solution ditto. We may indeed express the whole by a definite integral, but there does not seem to be much use in doing so, as there will be all the labour of finding out *its* solutions, and they are what we now obtain from the differential equations.

Let E_1 and E_2 be the E's of the inward and outward waves. Their equations are

$$E_1 = -(a/2q)W_a'Ue, \quad . \quad . \quad . \quad . \quad (377)$$

$$E_2 = -(a/2q)WU_a'e; \quad . \quad . \quad . \quad . \quad (378)$$

where U and W are given by (309), the accent means differentiation to r, and the suffix indicates the value at $r=a$. To prove these, it is sufficient to observe that U and W involve ϵ^{qr} and ϵ^{-qr} respectively, so that (377) expresses an inward and (378) an outward wave; and further that, by (310), we have

$$E_1 - E_2 = e \text{ at } r=a, \text{ always}; \quad . \quad . \quad . \quad (379)$$

which is the sole boundary condition at the surface of curl of **e**.

Expanding (377), we get

$$E_1 = \tfrac{1}{2}\left(\frac{a}{r}\right)^{\frac{1}{2}}\epsilon^{q(r-a)}(R+S)\left[1+\frac{3}{y}-\frac{3.5}{\underline{|2}\,y^2}+\frac{3^2.5.7}{\underline{|3}\,y^3}\right.$$
$$\left.-\frac{3^2.5.7.9}{\underline{|4}\,y^4}+\ldots\right]e, \quad . \quad (380)$$

where R+S is given by (309). Now e being zero before and constant after $t=0$, effect the integrations indicated by the inverse powers of p, and then turn t to t_1, where

$$vt_1 = vt + r - a.$$

The result is

$$\mathrm{E}_1=\tfrac{1}{2}e\left(\frac{a}{r}\right)^{\frac{1}{2}}\Big[1+3z_1-\frac{3.5}{\underline{|2}\,\underline{|2}}z_1^2+\frac{1^2.3^2.5.7}{\underline{|3}\,\underline{|3}}z_1^3+\frac{1^2.3^2.5^2.7.9}{\underline{|4}\,\underline{|4}}z_1^4+\dots$$
$$+\frac{a}{r}z_1\left(1+\frac{3}{\underline{|2}}z_1-\frac{3.5}{\underline{|2}\,\underline{|3}}z_1^2+\frac{1^2.3^2.5.7}{\underline{|3}\,\underline{|4}}z_1^3+\dots\right)$$
$$+\frac{1^2 3^2}{\underline{|2}}\frac{a^2}{r^2}z_1^2\left(\frac{1}{\underline{|2}}+\frac{3}{\underline{|3}}z_1-\frac{3.5}{\underline{|2}\,\underline{|4}}z_1^2+\frac{1^2.3^2.5.7}{\underline{|3}\,\underline{|5}}z_1^3-\dots\right)+\dots\Big], \quad (381)$$

the structure of which is sufficiently clear. Here $z_1=vt_1/8a$.

This formula, when $vt<a$, holds between $r=a$ and $r=a-vt$. But when $vt>a$ though $<2a$, it holds between $r=a$ and $vt-a$. Except within the limits named, it is only a partial solution.

70. As regards E_2, it may be obtained from (381) by the following changes. Change E_1 to $-\mathrm{E}_2$ on the left, and on the right change z_1 to $-z_2$, where

$$z_2=(vt+a-r)/8a.$$

It is therefore unnecessary to write out E_2. This E_2 formula will hold from $r=a$ to $r=vt+a$, when $vt<2a$; but after that, when the front of the return wave has passed $r=a$, it will only hold between $r=vt-a$ and $vt+a$.

71. Next to find E_3, the E in the cylinder when $vt>a$ and the solution is made up of two oppositely going waves, and E_4 the external E after $vt=2a$, when it is made up of two outward going waves. I have utterly failed to obtain intelligible results by uniting the primary waves with a reflected wave. But there is another method which is easier, and free from the obscurity which attends the simultaneous use of U and W. Thus, the equations of E_3 and E_4 are

$$\mathrm{E}_3=-(\pi/2q)^{\frac{1}{2}}a\mathrm{J}_{0r}\mathrm{W}_a'e, \quad \dots \dots \quad (382)$$

$$\mathrm{E}_4=-(\pi q^3/2)^{\frac{1}{2}}\tfrac{1}{2}a^2\left(\frac{2}{sa}\mathrm{J}_{1a}\right)\mathrm{W}.e, \quad \dots \quad (383)$$

by (367); and a necessity of their validity is the presence of two waves inside the cylinder, because of the use of J_0 and J_1; it is quite inadmissible to use J_0 when only one wave is in question, because $\mathrm{J}_{0r}=1$ when $r=0$, and being a differential operator in rising powers of p, the meaning of (382) is that we find E_3 at r by differentiations from E_3 at $r=0$; thus (382) only begins to be valid when $vt=a$.

To integrate (382), (383), it saves a little trouble to calculate the time-integrals of E_3 and E_4, say

$$\mathrm{A}_3=-p^{-1}\mathrm{E}_3, \quad \mathrm{A}_4=-p^{-1}\mathrm{E}_4. \quad \dots \quad (384)$$

The results are

$$-A_3 = J_{0r} \cdot \frac{e}{v}(v^2t^2 - a^2)^{\frac{1}{2}}, \quad . \quad . \quad . \quad . \quad . \quad . \quad (385)$$

$$A_4 = \left(\frac{2}{sa} J_{1a}\right)\frac{a^2 e}{2v}(v^2t^2 - r^2)^{-\frac{1}{2}}. \quad . \quad . \quad . \quad (386)$$

From these derive E_3 and E_4 by time-differentiation, and H_3, H_4 by space-differentiation, according to

$$\text{curl } \mathbf{A} = \mu \mathbf{H}, \quad \text{or } H = -\frac{1}{\mu}\frac{dA}{dr}. \quad . \quad . \quad . \quad (387)$$

We see that the value of E_3 at the axis, say E_0, is

$$E_0 = evt(v^2t^2 - a^2)^{-\frac{1}{2}}; \quad . \quad . \quad . \quad . \quad . \quad . \quad (388)$$

and by performing the operation J_{0r} in (385) we produce, if $u = (v^2t^2 - a^2)^{\frac{1}{2}}$,

$$-A_3 = \frac{e}{v}\left[u + \frac{r^2}{2^2}\left(\frac{1}{u} - \frac{v^2t^2}{u^3}\right) + \frac{3r^4}{2^2 4^2}\left(-\frac{1}{u^3} + \frac{6v^2t^2}{u^5} - \frac{5v^4t^4}{u^7}\right)\right.$$
$$\left. + \frac{45r^6}{2^2 4^2 6^2}\left(\frac{1}{u^5} - \frac{15v^2t^2}{u^7} + \frac{35v^4t^4}{u^9} - \frac{21v^6t^6}{u^{11}}\right) + \ldots\right]; \quad (389)$$

from which we derive

$$E_3 = \frac{evt}{u}\left[1 + \frac{3a^2r^2}{4u^4} + \frac{15a^2r^4}{2^2 4^2 u^8}(3a^2 + 4v^2t^2)\right.$$
$$\left. + \frac{5.7.9a^2r^6}{2^2.4^2.6^2u^{12}}(5a^4 + 20v^2t^2a^2 + 8v^4t^4) + \ldots\right], \quad (390)$$

These formulæ commence to operate when $vt = a$ at the axis and when $vt = a + r$ at any point $r < a$, and continue in operation for ever after.

72. Lastly, perform the operation $(2/sa)J_{1a}$ in (386), and we obtain

$$A_4 = \frac{a^2 e}{2v}\left[\frac{1}{u} + \frac{a^2}{8}\left(-\frac{1}{u^3} + \frac{3v^2t^2}{u^5}\right) + \frac{a^4}{64}\left(\frac{3}{u^5} - \frac{30v^2t^2}{u^7} + \frac{35v^4t^4}{u^9}\right)\right.$$
$$\left. + \frac{45a^6}{4.36.64}\left(-\frac{5}{u^7} + \frac{135v^2t^2}{u^9} - \frac{315v^4t^4}{u^{11}} + \frac{231v^6t^6}{u^{13}}\right) + \ldots\right] (391)$$

from which we derive

$$E_4 = \frac{a^2 evt}{2u^3}\left[1 + \frac{3a^2}{8u^4}(2v^2t^2 + 3r^2) + \frac{5a^4}{64u^8}(8v^4t^4 + 40v^2t^2r^2 + 15r^4)\right.$$
$$\left. + \frac{45a^6}{4.36.64u^{12}}(112v^6t^6 + 1176v^4t^4r^2 + 1470v^2t^2r^4 + 245r^6) + \ldots\right] (39$$

These begin to operate at $r=a$ when $vt=2a$; and later. the range is from $r=a$ to $r=vt-a$.

This completes the mathematical work. As a check upon the accuracy, we may test satisfaction of differential equations, and of the initial condition, and that the four solutions join together with the proper discontinuities.

73. The following is a general description of the manner of establishing the steady flux. We put on e in the cylinder when $t=0$. The first effect inside is $E_1=\frac{1}{2}e$ at the surface and $H_1=E_1/\mu v$. This primary disturbance runs in to the axis at speed v, varying at its front inversely as the square root of the distance from the axis, thus producing a momentary infinity there. At this moment $t=a/v$, E_1 is also very great near the axis. In the meantime E_1 has been increasing generally all over the cylinder, so that, from being $\frac{1}{2}e$ initially at the boundary, it has risen to $\cdot 77e$, whilst the simultaneous value at $r=\frac{1}{2}a$ is about $\cdot 95e$.

Now consider E_3 within the cylinder, it being the natural continuation of E_1. The large values of E_1 near the axis subside with immense rapidity. But near the boundary E_1 still goes on increasing. The result is that when $vt=2a$, and the front of the return wave reaches the boundary, E_3 has fallen from ∞ to $1\cdot 154e$ at the axis; at $r=\frac{1}{2}a$ the value is $1\cdot 183e$; at $r=\frac{3}{4}a$ it is $1\cdot 237e$; and at the boundary the value has risen to $1\cdot 71e$, which is made up thus, $1\cdot 21e+\frac{1}{2}e$; the first of these being the value just before the front of the return wave arrives, the second part the sudden increase due to the wave-front. E_3 is now a minimum at the axis and rises towards the wave-front, the greater part of the rise being near the wave-front.

Thirdly, go back to $t=0$ and consider the outward wave. First, $E_2=-\frac{1}{2}e$ at $r=a$. This runs out at speed v, varying at the front inversely as $r^{\frac{1}{2}}$. As it does so, the E_2 that succeeds rises, that is, is less negative. Thus when $vt=a$, and the front has got to $r=2a$, the values of E_2 are $-\cdot 232e$ at $r=a$ and $-\cdot 353e$ at $r=2a$. Still later, as this wave forms fully, its hinder part becomes positive. Thus, when fully formed, with front at $r=3a$, we have $E_2=-\cdot 288e$ at $r=3a$; $-\cdot 145e$ at $r=2a$; and $\cdot 21e$ at $r=a$. This is at the moment when the return wave reaches the boundary, as already described.

The subsequent history is that the wave E_2 moves out to infinity, being negative at its front and positive at its back, where there is a sudden rise due to the return wave E_4, behind which there is a rapid fall in E_4, not a discontinuity, but the continuation of the before-mentioned rapid fall in E_3 near its front. The subsidence to the steady state in the cylinder and outside is very rapid when the front of E_4 has moved well out. Thus, when $vt=5a$, we have $E_3=1{\cdot}022e$ at $r=a$, and of course, just outside, we have $E_4={\cdot}022e$; and when $vt=10a$, we have $E_3=1{\cdot}005e$, $E_4={\cdot}005e$, at $r=a$.

As regards H, starting when $t=0$ with the value $e/2\mu v$ at $r=a$ only, at the front of the inward or outward wave it is $E=\pm\mu v H$ as usual. It is positive in the cylinder at first and then changes to negative. Outside, it is first positive for a short time and then negative for ever after.

74. We can now see fully why tho solution for a filament e_0 of e can *not* be employed to build up more complex solutions in general, whilst that for a filament f_0 of curl e *can* be so employed. For, in the latter case, the disturbances come, *ab initio*, from the axis, because the lines of curl e are the sources of disturbance, and they become a single line at the axis. But in the former case it is not the body of the filament, but its surface only, that is the real source, however small the filament may be, producing first **E** negative (or against e) just outside the filament, and immediately after **E** positive. Now when the diameter of the filament is indefinitely reduced, we lose sight altogether of the preliminary negative electric and positive magnetic force, because their duration becomes infinitely small, and our solutions (372) show only the subsequent state of positive electric and negative magnetic force during the settling down to the final state, but not its real commencement, viz. at the front of the wave.

75. The occurrence of momentary infinite values of **E** or of **H**, in problems concerning spherical and cylindrical electromagnetic waves, is physically suggestive. By means of a proper convergence to a point or an axis, we should be able to disrupt the strongest dielectric, starting with a weak field, and then discharging it. Although it is impossible to realize the particular arrangements of our solutions, yet it might be practicable to obtain similar results in other ways*.

* If we wish the solution for an infinitely long cylinder to be quite unaltered, when of finite length l, let at $z=0$ and $z=l$ infinitely conducting barriers be placed. Owing to the displacement terminating upon them perpendicularly, and the magnetic force being tangential, no alteration is required. Then, on taking off the impressed force, we obtain the result of the discharge of a condenser consisting of two parallel plates of no resistance, charged in a certain portion only; or, by integration, charged in any manner.

To abolish the momentary infinity at the axis, in the text, substitute for the surface distribution of curl of e a distribution in a thin layer. The infinity will be replaced by a large finite value, without other

It may be remarked that the solution worked out for an infinitely long cylinder of longitudinal **e** is also, to a certain extent, the solution for a cylinder of finite length. If, for instance, the length is $2l$, and the radius a, disturbances from the extreme terminal lines of **f** (or curl **e**) only reach the centre of the axis after the time $(a^2+l^2)^{\frac{1}{2}}/v$, whilst from the equatorial line of **f** the time taken is a/v, which may be only a little less, or very greatly less, according as l/a is small or large. If large, it is clear that the solutions for **E** and **H** in the central parts of the cylinder are not only identical with those for an infinitely long cylinder until disturbances arrive from its ends, but are not much different afterwards.

76. *Cylindrical surface of longitudinal* **f**, *a function of* θ *and* t.—When there is no variation with θ, the only Bessel functions concerned are J_0 and J_1. The extension of the vibratory solutions to include variation of the impressed force or its curl as $\cos\theta$, $\cos 2\theta$, &c. is so easily made that it would be inexcusable to overlook it. Two leading cases will be very briefly considered. Let the curl of the impressed force be wholly upon the surface of a cylinder of radius a, longitudinally directed, and be a function of t and θ, its tensor being f, the measure of the surface-density. **H** is also longitudinal of course, whilst **E** has two components, circular E and radial F. The connexions are

$$\left.\begin{aligned} -\frac{d\mathrm{H}}{dr} = cp\mathrm{E}, \quad \frac{1}{r}\frac{d\mathrm{H}}{d\theta} = cp\mathrm{F}, \\ \frac{1}{r}\frac{d}{dr}r\mathrm{E} - \frac{1}{r}\frac{d\mathrm{F}}{d\theta} = -\mu p\mathrm{H}, \end{aligned}\right\} \quad . \quad . \quad . \quad (393)$$

from which the characteristic of H is

$$\frac{1}{r}\frac{d}{dr}r\frac{d\mathrm{H}}{dr} + \left(s^2 - \frac{m^2}{r^2}\right)\mathrm{H} = 0, \quad . \quad . \quad . \quad . \quad . \quad (394)$$

if $s^2 = -p^2/v^2$ and $m^2 = -d^2/d\theta^2$. Consequently

$$\mathrm{H} = (\mathrm{J}_{m} - y\mathrm{G}_{mr}) \cos m\theta \times \text{function of } t \quad . \quad . \quad (395)$$

when m^2 is constant, and the E/H operator is

$$\frac{\mathrm{E}}{\mathrm{H}} = -\frac{1}{cp}\frac{\mathrm{J}'_{mr} - y\mathrm{G}'_{mr}}{\mathrm{J}_{mr} - y\mathrm{G}_{mr}}, \quad . \quad . \quad . \quad . \quad . \quad (396)$$

---

material change. Of course the theory above assumes that the dielectric does not break down. If it does, we change the problem, and have a conducting (or resisting) path, possibly with oscillations of great frequency, if the resistance be not too great, as Prof. Lodge believes to be the case in a lightning discharge.

if J_{mr} or $J_m(sr)$ is the mth Bessel function, and G_{mr} its companion, whilst the $'$ means d/dr.

The boundary condition is

$$E_1 = E_2 - f \quad \text{at} \quad r = a, \quad . \quad . \quad . \quad . \quad (397)$$

E_1 being the inside, E_2 the outside value of the force of the flux. Therefore, using (396) with $y=0$ inside, we obtain

$$H_a = \frac{J_{ma}(J_{ma} - yG_{ma})}{y(J_{ma}G'_{ma} - J'_{ma}G_{ma})} cpf$$

$$= \frac{axcp}{y} J_{ma}(J_{ma} - y\,G_{ma})f, \quad . \quad . \quad . \quad . \quad (398)$$

where x is a constant, being $\pi/2$ when $m=0$, according to (307), and always $\pi/2$ if G_m has the proper numerical factor to fix its size.

We see that when

$$f = f_0 \cos m\theta \cos nt$$

when f_0 is constant, the boundary H, and with it the whole external field, electric and magnetic, vanishes when

$$J_{ma} = 0.$$

If $m=0$, or there is no variation with θ, the impressed force may be circular, outside the cylinder, and varying as r^{-1}.

If $m=1$, the impressed force may be transverse, within the cylinder, and of uniform intensity.

77. *Conducting tube.* **e** *circular, a function of* θ *and* t.—This is merely chosen as the easiest extension of the last case. In it let there be two cylindrical surfaces of **f**, infinitely close together. They will cancel one another if equal and opposite, but if we fill up the space between them with a tube of conductance K per unit area, we get the case of **e** circular in the tube, **e** varying with θ and t, and produce a discontinuity in H (which is still longitudinal, of course). Let E_a be the common value of E just outside and inside the tube; $e + E_a$ is then the force of the flux in the substance of the tube, and

$$H_1 - H_2 = 4\pi K(e + E_a) \quad . \quad . \quad . \quad . \quad (399)$$

the discontinuity equation, leads, by the use of (396) and the conjugate property of J_m and G_m as standardized in the last paragraph, through

$$\left(\frac{H_1}{E_1} - \frac{H_2}{E_2} - 4\pi K\right)E^a = 4\pi Ke$$

to the equation of E_a, viz.

$$E_a = \frac{4\pi K e}{-4\pi K + \frac{2cpy}{\pi a}[J'_{ma}(J'_{ma} - yG'_{ma})]^{-1}}, \quad . \quad . \quad . \quad (400)$$

from which we see that it is $J'_{ma}=0$ that now makes the external field vanish.

78. This concludes my treatment of electromagnetic waves in relation to their sources, so far as a systematic arrangement and uniform method is concerned. Some cases of a more mixed character must be reserved. It is scarcely necessary to remark that all the dielectric solutions may be turned into others, by employing impressed magnetic instead of electric force. The hypothetical magnetic conductor is required to obtain full analogues of problems in which electric conductors occur.

August 10, 1888.

From the PHILOSOPHICAL MAGAZINE for January 1889.

The General Solution of Maxwell's *Electromagnetic Equations in a Homogeneous Isotropic Medium, especially in regard to the Derivation of special Solutions, and the Formulæ for Plane Waves.* *By* OLIVER HEAVISIDE.

1. *EQUATIONS of the Field.*—Although, from the difficulty of applying them to practical problems, general solutions frequently possess little practical value, yet they may be of sufficient importance to render their investigation desirable, and their applications examined as far as may be practicable. The first question here to be answered is this. Given the state of the whole electromagnetic field at a certain moment, in a homogeneous isotropic conducting dielectric medium, to deduce the state at any later time, arising from the initial state alone, without impressed forces.

The equations of the field are, if p stand for d/dt,

$$\text{curl}\ \mathbf{H} = (4\pi k + cp)\ \mathbf{E}, \qquad (1)$$

$$-\text{curl}\ \mathbf{E} = (4\pi g + \mu p)\ \mathbf{H}; \qquad (2)$$

the first being Maxwell's well-known equation defining electric current in terms of the magnetic force $\mathbf{H}$, k being the electric conductivity and $c/4\pi$ the electric permittivity (or permittance of a unit-cube condenser), and $\mathbf{E}$ the electric force; whilst the second is the equation introduced by me* as the proper companion to the former to make a complete system suitable for practical working, g being the magnetic conductivity and μ the magnetic inductivity. This second equation takes the place of the two equations

$$\left.\begin{aligned}\mathbf{E}&=-\dot{\mathbf{A}}-\nabla\Psi,\\ \mathrm{curl}\,\mathbf{A}&=\mu\mathbf{H},\end{aligned}\right\}\quad . \quad . \quad . \quad . \quad . \qquad (3)$$

of Maxwell, where $\mathbf{A}$ is the electromagnetic momentum at a point and Ψ the scalar electric potential. Thus Ψ and $\mathbf{A}$ are murdered, so to speak, with a great gain in definiteness and conciseness. As regards g, however, standing for a physically non-existent quality, such that the medium cannot support magnetic force without a dissipation of energy at the rate $g\mathbf{H}^2$ per unit volume, it is only retained for the sake of mathematical completeness, and on account of the singular telegraphic application in which electric conductivity is made to perform the functions of both the real k and the unreal g.

Let

$$\left.\begin{aligned}\rho_1&=4\pi k/2c, & \rho&=\rho_1+\rho_2, & v&=(\mu c)^{-\frac{1}{2}}.\\ \rho_2&=4\pi g/2\mu, & \sigma&=\rho_1-\rho_2,\end{aligned}\right\}\quad . \quad . \qquad (4)$$

The speed of propagation of all disturbances is v, and the attenuating effects due to the two conductivities depend upon ρ_1 and ρ_2, whilst σ determines the distortion due to conductivity.

2. *General Solutions.*—Let q^2 denote the operator

$$q^2=-(v\ \mathrm{curl})^2+\sigma^2; \quad . \quad . \quad . \quad . \quad . \qquad (5)$$

or, in full, when operating upon $\mathbf{E}$ for example,

$$q^2\mathbf{E}=v^2\nabla^2\mathbf{E}-v^2\nabla\,\mathrm{div}\,\mathbf{E}+\sigma^2\mathbf{E}. \quad . \quad . \quad . \qquad (6)$$

Now it may be easily found by ordinary "symbolical" work which it is not necessary to give, that, given $\mathbf{E}_0$, $\mathbf{H}_0$, the values of $\mathbf{E}$ and $\mathbf{H}$ when $t=0$, and satisfying (1) and (2), those at time t later are given by

$$\left.\begin{aligned}\mathbf{E}&=\epsilon^{-\rho t}\left[\left(\cosh qt-\frac{\sigma}{q}\sinh qt\right)\mathbf{E}_0+\frac{\sinh qt}{q}\cdot\frac{\mathrm{curl}\,\mathbf{H}_0}{c}\right],\\ \mathbf{H}&=\epsilon^{-\rho t}\left[\left(\cosh qt+\frac{\sigma}{q}\sinh qt\right)\mathbf{H}_0-\frac{\sinh qt}{q}\cdot\frac{\mathrm{curl}\,\mathbf{E}_0}{\mu}\right].\end{aligned}\right\}(7)$$

* "Electromagnetic Induction and its Propagation," the 'Electrician,' January 3, 1885, and later.

A sufficient proof is the satisfaction of the equations (1), (2), and of the two initial conditions.

An alternative form of (7) is

$$\left.\begin{aligned}\mathbf{E}&=\epsilon^{-\rho t}\left[\cosh qt+\frac{\sinh qt}{q}(p+\rho)\right]\mathbf{E}_0,\\ \mathbf{H}&=\epsilon^{-\rho t}\left[\cosh qt+\frac{\sinh qt}{q}(p+\rho)\right]\mathbf{H}_0,\end{aligned}\right\} \quad . \quad . \quad (7\,a)$$

showing the derivation of $\mathbf{E}$ from $\mathbf{E}_0$ and $p\mathbf{E}_0$ in precisely the same way as $\mathbf{H}$ from $\mathbf{H}_0$ and $p\mathbf{H}_0$. In this form of solution the initial values of $p\mathbf{E}_0$ and $p\mathbf{H}_0$ occur. But they are not arbitrary, being connected by equations (1), (2). The form (7) is much more convenient, involving only $\mathbf{E}_0$ and $\mathbf{H}_0$ as functions of position, although (7 *a*) looks simpler. The form (7) is also the more useful for interpretations and derivations.

If, then, $\mathbf{E}_0$ and $\mathbf{H}_0$ be given as continuous functions admitting of the performance of the differentiations involved in the functions of q^2, (7) will give the required solutions. The original field should therefore be a real one, not involving discontinuities. We shall now consider special cases.

3. *Persistence or Subsidence of Polar Fields.*—We see immediately by (7) that the $\mathbf{E}$ resulting from $\mathbf{H}_0$ depends solely upon its curl, or on the initial electric current, and, similarly, that the $\mathbf{H}$ due to $\mathbf{E}_0$ depends solely upon its curl, or on the magnetic current. Notice also that the displacement due to $\mathbf{H}_0$ is related to $\mathbf{H}_0$ in the same way as the induction $\div -4\pi$ due to $\mathbf{E}_0$ is related to $\mathbf{E}_0$. Or, if it be the electric and magnetic currents that are considered, the displacement due to electric current is related to it in the same way as the induction $\div 4\pi$ due to magnetic current is related to it.

Observe also that in passing from the $\mathbf{E}$ due to $\mathbf{E}_0$ to the $\mathbf{H}$ due to $\mathbf{H}_0$ the sign of σ is changed.

By (7) a distribution of $\mathbf{H}_0$ which has no curl, or a polar magnetic field, does not, in subsiding, generate electric force; and, similarly, a polar electric field does not, in subsiding, generate magnetic force. Let then $\mathbf{E}_0$ and $\mathbf{H}_0$ be polar fields, in the first place. Then, by (5),

$$q^2=\sigma^2,$$

that is, a constant, and using this in (7) we reduce the general solutions to

$$\mathbf{E}=\mathbf{E}_0\epsilon^{-2\rho_1 t}, \qquad \mathbf{H}=\mathbf{H}_0\epsilon^{-2\rho_2 t}. \quad . \quad . \quad . \quad . \quad (8)$$

The subsidence of the electric field requires electric conductivity, that of the magnetic field requires magnetic conductivity;

but the two phenomena are wholly independent. The first of (8) is equivalent to Maxwell's solution*. The second is its magnetic analogue.

As, in the first case, there must be initial electrification, so in the second, there should be "magnetification," its volume-density to be measured by the divergence of the induction $\div 4\pi$. Now the induction can have no divergence. But it might have, if g existed.

There is no true electric current during the subsidence of $\mathbf{E}_0$, and there would be no true magnetic current during the subsidence of $\mathbf{H}_0$. In both cases the energy is frictionally dissipated on the spot, or there is no transfer of energy†. The application of (8) will be extended later.

4. *Purely Solenoidal Fields.*—By a purely solenoidal field I mean one which has no divergence anywhere. Any field vanishing at infinity may be uniquely divided into two fields, one of which is polar, the other solenoidal; the proof thereof resting upon Sir W. Thomson's well-known theorem of Determinancy. Now we know exactly what happens to the polar fields. Therefore dismiss them, and let $\mathbf{E}_0$ and $\mathbf{H}_0$ be solenoidal. Then

$$q^2 = v^2\nabla^2 + \sigma^2, \quad . \quad . \quad . \quad . \quad . \quad . \quad . \quad (9)$$

where ∇^2 is the usual Laplacean operator. Of course $\cosh qt$ and $q^{-1}\sinh qt$ are rational functions of q^2, so that if the differentiations are possible we shall obtain the solutions out of (7).

5. *Non-distortional Cases.*—Let the subsidence-rates of the polar electric and magnetic fields be equal. We then have

$$\sigma = 0, \qquad \left.\begin{array}{c} q^2 = -(v \text{ curl})^2, \\ \rho = 4\pi k/c = 4\pi g/\mu, \end{array}\right\} \quad . \quad . \quad . \quad (10)$$

in the solutions (7). The fields change in precisely the same manner as if the medium were nonconducting, as regards the relative values at different places; that is, there is no distortion due to the conductivities; but there is a uniform subsidence all over brought in by them‡, expressed by the factor $\epsilon^{-\rho t}$. This property I have explained by showing the opposite nature of the tails left behind by a travelling plane wave according as σ is + or −.

* Vol. i. chap. x. art. 325, equation (4).

† This is of course obvious without any reference to Poynting's formula. The only other simple case of no transfer of energy which had been noticed before that formula is that of conduction-current, kept up by impressed force so distributed as to require no polar force to supplement it.

‡ "Electromagnetic Waves," Part I. § 7, Phil. Mag. February 1888.

The above applies to a homogeneous medium. But, if in

$$\left.\begin{aligned}\text{curl}\,(\mathbf{H}-\mathbf{h})&=(4\pi k+cp)\mathbf{E},\\ \text{curl}\,(\mathbf{e}-\mathbf{E})&=(4\pi g+\mu p)\mathbf{H},\end{aligned}\right\}\quad\begin{aligned}(1\,a)\\(2\,a)\end{aligned}$$

differing from (1), (2) only in the introduction of impressed forces **e** and **h**, we write

$$(\mathbf{H},\mathbf{h},\mathbf{E},\mathbf{e})=(\mathbf{H}_1,\mathbf{h}_1,\mathbf{E}_1,\mathbf{e}_1)\epsilon^{-\rho t},$$

we reduce them to

$$\left.\begin{aligned}\text{curl}\,(\mathbf{H}_1-\mathbf{h}_1)&=c(\sigma+p)\mathbf{E}_1\\ \text{curl}\,(\mathbf{e}_1-\mathbf{E}_1)&=\mu(-\sigma+p)\mathbf{H}_1\end{aligned}\right\}\quad . \quad . \quad . \quad (11)$$

and these, if $\sigma=0$, are the equations of a nonconducting dielectric. That is,

$$\rho=4\pi k/c=4\pi g/\mu=\text{constant}$$

is the required condition. Therefore c and μ may vary anyhow, independently, provided k and g vary similarly*. The impressed forces should subside according to $\epsilon^{-\rho t}$, in order to preserve similarity to the phenomena in a nonconducting dielectric.

Observe that there will be tailing now, on account of the variability of $(\mu/c)^{\frac{1}{2}}$ or μv. That is, there are reflexions and refractions due to change of medium. The peculiarity is that they are of the same nature with as without conductivity.

6. *First Special Case.*—A special case of (11) is given by taking $\mu=0$ and $g=0$; that is, a real conducting dielectric possessing no magnetic inductivity, in which k/c is constant. If the initial field be polar, then

$$\mathbf{E}=\mathbf{E}_0\epsilon^{-\rho t},\qquad \mathbf{H}=0.\quad . \quad . \quad . \quad (12)$$

This extension of Maxwell's before-mentioned solution I have given before, and also the extension to any initial field, and the inclusion of impressed forces†. The theory of the result has considerable light now thrown upon it.

If the initial field be arbitrary, the solenoidal part of the flux displacement disappears instantly, therefore (12) is the solution, provided $\mathbf{E}_0$ means the polar part of the initial field; that is, $\mathbf{E}_0$ must have no curl, and the flux $c\mathbf{E}_0/4\pi$ must have the same divergence as the arbitrarily given displacement.

Now an impressed force **e** produces a solenoidal flux only. Therefore it produces its full effect and sets up the appropriate

* In § 4 of the article referred to in the last footnote the property was described only in reference to a homogeneous medium.

† "Electromagnetic Induction," 'Electrician,' December 18, 1885, and January 1, 1886.

steady flux instantaneously; and all variations of **e** in time and in space are kept time to without lag by the conduction-current in spite of the electric displacement.

This property is seemingly completely at variance with ideas founded upon the retardation usually associated with combinations of resistances and condensers. But, being a special case of the nondistortional theory, we can now understand it. For suppose we start with a nonconducting dielectric, and put on **e** uniform within a spherical portion thereof, and send out an electromagnetic wave to infinity and set up the steady flux. On now removing **e**, we send out another wave to infinity, and the flux vanishes. Now make the medium conducting, with both conductivities balanced, as in (10). Starting with the same steady flux, its vanishing will take place in the same manner precisely, but with an attenuation factor $\epsilon^{-\rho t}$. Now gradually reduce g and μ at the same time, in the same ratio. The vanishing of the flux will take place faster and faster, and in the limit, when both μ and g are zero, will take place instantly, not by subsidence, but by instantaneous transference to an infinite distance when the impressed force is removed, owing to v being made infinite.

7. *Second Special Case.*—There is clearly a similar property when $k=0$ and $c=0$, that is, in a medium possessing magnetic inductivity and conductivity, but deprived of the electric correspondences. Thus, when g/μ is constant, the solution due to any polar field $\mathbf{H}_0$ is

$$\mathbf{H}=\mathbf{H}_0\epsilon^{-\rho t}, \qquad \mathbf{E}=0; \quad . \quad . \quad . \quad . \quad (13)$$

wherein $\rho=4\pi g/\mu$. But a solenoidal field of $\mu\mathbf{H}$ disappears at once, by instantaneous transference to infinity. Thus any varying impressed force **h** is accompanied without delay by the corresponding steady flux, the magnetic induction.

When the inertia associated with μ is considered the result is rather striking and difficult to understand. It appears, however, to belong to the same class of (theoretical) phenomena as the following. If a coil in which there is an electric current be instantaneously shunted on to a second coil in which there is no current, then, according to Maxwell, the first coil instantly loses current and the second gains it, in such a way as to keep the momentum unchanged. Now we cannot set up a current in a coil instantly, so that we have a contradiction. But the disagreement admits of easy reconciliation. We cannot set up current instantly with a finite impressed force, but if it be infinite we can. In the case of the coils there is an electromotive impulse, or infinite electromotive force acting for an infinitely short time, when the coils are connected, with corresponding instantaneous changes in their momenta. A loss of energy is involved.

It is scarcely necessary to remark that the true physical theory involves other considerations on account of the dielectric not being infinitely elastive and on account of diffusion in the wires; so that we have sparking and very rapid vibrations in the dielectric. The energy which is not wasted in the spark, and which would go out to infinity were there no conducting obstacles, is probably all wasted practically in the heat of conduction-currents in them.

8. *Impressed Forces.*—Given initially $\mathbf{E}_0$ and $\mathbf{H}_0$, we know that the diverging parts must either remain constant or subside, and are, in a manner, self-contained; but the solenoidal fields, which would give rise to waves, may be kept from changing by means of impressed forces $\mathbf{e}_0$ and $\mathbf{h}_0$. Thus let $\mathbf{E}_0$ and $\mathbf{H}_0$ be solenoidal. To keep them steady we have, in equations (1), (2), to get rid of $p\mathbf{E}$ and $p\mathbf{H}$. Thus

$$\left.\begin{aligned}\text{curl}\,(\mathbf{H}_0-\mathbf{h}_0)&=4\pi k\mathbf{E}_0,\\ \text{curl}\,(\mathbf{e}_0-\mathbf{E}_0)&=4\pi g\mathbf{H}_0,\end{aligned}\right\}\quad . \quad . \quad . \quad . \quad (14)$$

are the equations of steady fields $\mathbf{E}_0$ and $\mathbf{H}_0$, these being the forces of the fluxes. Or

$$\left.\begin{aligned}\text{curl}\,\mathbf{h}_0&=\text{curl}\,\mathbf{H}_0-4\pi k\mathbf{E}_0,\\ \text{curl}\,\mathbf{e}_0&=\text{curl}\,\mathbf{E}_0+4\pi g\mathbf{H}_0,\end{aligned}\right\}\quad . \quad . \quad . \quad (14\,a)$$

gives the curls of the required impressed forces in terms of the given fluxes, and any impressed forces having these curls will suffice.

Now, on the sudden removal of $\mathbf{e}_0$, $\mathbf{h}_0$, the forces $\mathbf{E}_0$, $\mathbf{H}_0$, which had hitherto been the forces of the fluxes, become, instantaneously, the forces of the field as well. That is, the fluxes themselves do not change suddenly, except in such a case as a tangential discontinuity in a flux produced at a surface of curl of impressed force when, at the surface itself, the mean value will be immediately assumed on removal of the impressed force. We know, therefore, the effects due to certain distributions of impressed force when we know the result of leaving the corresponding fluxes to themselves without impressed force. It is, however, the converse of this that is practically useful, viz. to find the result of leaving the fluxes without impressed force by solving the problem of the establishment of the steady fluxes when the impressed forces are suddenly started; because this problem can often be attacked in a comparatively simple manner, requiring only investigation of the appropriate functions to suit the surfaces of curl of the impressed forces. The remarks in this paragraph are not limited to homogeneity and isotropy.

9. *Solutions for Plane Waves.*—If we take z normal to the plane of the waves, we may suppose that both **E** and **H** have x and y components. This is, however, a wholly unnecessary mathematical complication, and it is sufficient to suppose that **E** is everywhere parallel to the x-axis and **H** to the y-axis. The specification of an initial state is therefore E_0, H_0, the tensors of **E** and **H**, given as functions of z; and the equations of motion (1), (2) become

$$\left.\begin{aligned} -\frac{dH}{dz}&=(4\pi k+cp)E, \\ -\frac{dE}{dz}&=(4\pi g+\mu p)H. \end{aligned}\right\} \quad . \quad . \quad . \quad . \quad (15)$$

Now the operator q^2 in (5) becomes

$$q^2=v^2\nabla^2+\sigma^2\,; \quad . \quad . \quad . \quad . \quad . \quad . \quad (16)$$

where by ∇ we may now understand d/dz simply. Therefore, by (7), the solutions of (15) are

$$\left.\begin{aligned} E&=\epsilon^{-\rho t}\left[\left(\cosh qt-\frac{\sigma}{q}\sinh qt\right)E_0-\frac{\sinh qt}{q}\frac{\nabla}{c}H_0\right]. \\ H&=\epsilon^{-\rho t}\left[\left(\cosh qt+\frac{\sigma}{q}\sinh qt\right)H_0-\frac{\sinh qt}{q}\frac{\nabla}{\mu}E_0\right]. \end{aligned}\right\} (17)$$

When the initial states are such as $a\epsilon^{bz}$, or $a\cos bz$, the realization is immediate, requiring only a special meaning to be given to q in (17). But with more useful functions as $a\epsilon^{-bz^2}$, &c., &c., there is much work to be performed in effecting the differentiations, whilst the method fails altogether if the initial distribution is discontinuous.

But we may notice usefully that when E_0 and H_0 are constants the solutions are

$$E=\epsilon^{-2\rho_1 t}E_0, \qquad H=\epsilon^{-2\rho_2 t}H_0, \quad . \quad . \quad . \quad (18)$$

which are quite independent of one another. Further, since disturbances travel at speed v, (18) represents the solutions in any region in which E_0 and H_0 are constant, from $t=0$ up to the later time when a disturbance arrives from the nearest plane at which E_0 or H_0 varies.

10. *Fourier Integrals.*—Now transform (17) to Fourier integrals. We have Fourier's theorem,

$$f(z)=\frac{1}{\pi}\int_0^\infty\int_{-\infty}^\infty f(a)\cos m(z-a)\,dm\,da, \quad . \quad . \quad (19)$$

and therefore

$$\phi(\nabla^2)f(z) = \frac{1}{\pi}\int_0^\infty\int_{-\infty}^\infty f(a)\phi(-m^2)\cos m(z-a)\,dm\,da; \quad (20)$$

applying which to (17) we obtain

$$\left.\begin{aligned} E &= \frac{\epsilon^{-\rho t}}{\pi}\int_0^\infty\int_{-\infty}^\infty dm\,da\Big[E_0\cos m(z-a)\Big(\cosh - \frac{\sigma}{q}\sinh\Big)qt \\ &\qquad + \frac{H_0}{c}m\sin m(z-a)\frac{\sinh qt}{q}\Big]. \\ H &= \frac{\epsilon^{-\rho t}}{\pi}\int_0^\infty\int_{-\infty}^\infty dm\,da\Big[H_0\cos m(z-a)\Big(\cosh + \frac{\sigma}{q}\sinh\Big)qt \\ &\qquad + \frac{E_0}{\mu}m\sin m(z-a)\frac{\sinh qt}{q}\Big]. \end{aligned}\right\} \quad (21)$$

in which, by (16),

$$q^2 = \sigma^2 - m^2v^2, \quad . \quad . \quad . \quad . \quad . \quad . \quad (22)$$

and E_0, H_0 are to be expressed as functions of a, whilst E and H belong to z. Discontinuities are now attackable.

The integrations with respect to m may be effected. In fact, I have done it in three different ways. First by finding the effect produced by impressed force. Secondly, by an analogous method applied to (17), transforming the differentiations to integrations. Thirdly, by direct integration of (21); this is the most difficult of all. The first method was given in a recent paper* ; a short statement of the other two methods follows.

11. *Transformation of* (17).—In (17) we naturally consider the functions of qt to be expanded in rising powers of q^2, and therefore of ∇^2, leading to differentiations to be performed upon the initial states. But if we expand them in descending powers of ∇, we substitute integrations, and can apply them to discontinuous initial distribution.

The following are the expansions required :—

$$\left.\begin{aligned} \frac{\epsilon^{qt}}{q} &= \frac{1}{v\nabla}\Big[U_0 + U_1\Big(\frac{\sigma^2 t}{2v\nabla}\Big) + \frac{U_2}{\underline{|2}}\Big(\frac{\sigma^2 t}{2v\nabla}\Big)^2 + \frac{U_3}{\underline{|3}}\Big(\frac{\sigma^2 t}{2v\nabla}\Big)^3 + \ldots\Big], \\ \epsilon^{qt} &= U_0 + U_0\Big(\frac{\sigma^2 t}{2v\nabla}\Big) + \frac{U_1}{\underline{|2}}\Big(\frac{\sigma^2 t}{2v\nabla}\Big)^2 + \frac{U_2}{\underline{|3}}\Big(\frac{\sigma^2 t}{2v\nabla}\Big)^3 + \ldots, \end{aligned}\right\} \quad (23)$$

* "Electromagnetic Waves," Phil. Mag. October 1888.

where the U's are functions of $(v\nabla t)^{-1}$ given by

$$\left.\begin{aligned} &U_0=\epsilon^{vt\nabla}, \quad U_1=\epsilon^{vt\nabla}\left(1-\frac{1}{vt\nabla}\right), \quad U_2=\epsilon^{vt\nabla}\left(1-\frac{3}{vt\nabla}+\frac{3}{(vt\nabla)^2}\right), \\ &U_r=\epsilon^{vt\nabla}\left[1-\frac{r(r+1)}{2vt\nabla}+\frac{r(r^2-1^2)(r+2)}{2\,.\,4\,.\,(vt\nabla)^2}\right. \\ &\qquad\qquad\left.-\frac{r(r^2-1^2)(r^2-2^2)(r+3)}{2\,.\,4\,.\,6\,(vt\nabla)^3}+\ldots\right]; \end{aligned}\right\} \quad (24)$$

being in fact identically the same functions of $vt\nabla$ as those of r which occur in the investigation of spherical waves.

Arranged in powers of $s=\sigma/v\nabla$, we have

$$\left.\begin{aligned} \frac{\epsilon^{qt}}{q} &= \frac{\epsilon^{vt\nabla}}{v\nabla}(1+sg_1+s^2g_2+\ldots), \\ \epsilon^{qt} &= \epsilon^{vt\nabla}(1+sh_1+s^2h_2+\ldots), \end{aligned}\right\} \quad . \quad . \quad . \quad (25)$$

where

$$\left.\begin{aligned} &g_1=\frac{\sigma t}{2}, \quad g_2=-\tfrac{1}{2}+\frac{(\sigma t)^2}{2\,.\,4}, \quad g_3=-\tfrac{3}{4}\frac{\sigma t}{2}+\frac{(\sigma t)^3}{2\,.\,4\,.\,6}, \\ &g_4=\tfrac{3}{8}-\frac{(\sigma t)^2}{2\,.\,4}+\frac{(\sigma t)^4}{2\,.\,4\,.\,6\,.\,8}, \quad g_5=\tfrac{5}{8}\frac{\sigma t}{2}-\tfrac{5}{4}\frac{(\sigma t)^3}{2\,.\,4\,.\,6}+\frac{(\sigma t)^5}{2\,.\,4\,.\,6\,.\,8\,.\,10}, \\ &g_6=-\tfrac{5}{16}+\tfrac{15}{16}\frac{(\sigma t)^2}{2\,.\,4}-\tfrac{5}{2}\frac{(\sigma t)^4}{2\,.\,4\,.\,6\,.\,8}+\frac{(\sigma t)^6}{2\,.\,4\ldots\,.\,12}; \end{aligned}\right\}$$

$$\left.\begin{aligned} &h_1=\frac{\sigma t}{2}, \quad h_2=\frac{(\sigma t)^2}{2^2\lfloor 2}, \quad h_3=\frac{1}{2^2\lfloor 2}\left(-\sigma t+\frac{(\sigma t)^3}{2\,.\,3}\right), \\ &h_4=\frac{1}{2^3\lfloor 3}\left(-3(\sigma t)^2+\frac{(\sigma t)^4}{2\,.\,4}\right), \; h_5=\frac{1}{2^3\lfloor 3}\left(\sigma t-\frac{6(\sigma t)^3}{2\,.\,4}+\frac{(\sigma t)^5}{2^2.4.5}\right), \\ &h_6=\frac{1}{2^4\lfloor 4}\left(15(\sigma t)^2-\frac{10(\sigma t)^4}{2\,.\,5}+\frac{(\sigma t)^6}{2^2.5.6}\right), \\ &h_7=\frac{1}{2^4\lfloor 4}\left(-15\sigma t+\frac{45(\sigma t)^3}{2\,.\,5}-\frac{15(\sigma t)^5}{2^2.5.6}+\frac{(\sigma t)^7}{2^2.5.6.7}\right), \end{aligned}\right\} \quad (2$$

The following properties of the g's and h's are useful. Understanding that g_0 and h_0 are unity, we have

$$\left.\begin{aligned} &g_r+\sigma t g_{r+1}+\frac{(\sigma t)^2}{\lfloor 2}g_{r+2}+\ldots=0 \text{ when } r \text{ is odd}, \\ &\text{and when } r \text{ is even}, \qquad =1\,.\,3\,.\,5\ldots(r-1)(-1)^{\frac{r}{2}}\frac{J_{\frac{r}{2}}(\sigma ti)}{(\sigma ti)^{\frac{r}{2}}}. \\ &\text{except } r=0, \text{ when} \qquad =J_0(\sigma ti). \end{aligned}\right\} \quad ($$

$$h_r + \sigma t h_{r+1} + \frac{(\sigma t)^2}{\underline{|2}} h_{r+2} + \ldots = (-1)^{r+1}\sigma t \cdot \frac{J_r(\sigma t i)}{(\sigma t i)^r}, \quad . \quad (29)$$

except when $r=0$, which case is not wanted. Now if

$$\epsilon^{qt}\left(1+\frac{\sigma}{q}\right) = \epsilon^{vt\nabla}(1+sf_1+s^2f_2+\ldots), \quad . \quad . \quad (30)$$

the f's* will be given by (25), viz.

$$f_0=1, \quad f_1=g_0+h_1, \quad f_2=g_1+h_2, \text{ \&c.}; \quad . \quad . \quad . \quad (31)$$

and the properties of the f's corresponding to (28), (29) are

$$\left.\begin{aligned} f_r + \sigma t f_{r+1} + \frac{(\sigma t)^2}{\underline{|2}} f_{r+2} + \ldots &= \epsilon^{\sigma t} \text{ when } r=0, \\ &= 0 \text{ when } r \text{ is even, except } 0; \end{aligned}\right\} \quad (32)$$

and

$$\pm 1 \cdot 3 \cdot 5 \ldots (r-2) \; \frac{J_{\frac{r-1}{2}}(\sigma t i) - i J_{\frac{r+1}{2}}(\sigma t i)}{(\sigma t i)^{\frac{r-1}{2}}}; \quad . \quad . \quad (33)$$

when r is odd, with the + sign for $r=1$, 5 9,..., and the − sign for the rest. The first case in (32), of $r=0$, is very important. But in case $r=1$, the coefficient in (33) is +1; thus,

$$= (J_0 - iJ_1)(\sigma t i).$$

12. *Special Initial States.*—Now let there be an initial distribution of H_0 only, so that, by (17),

$$\left.\begin{aligned} H &= \epsilon^{-\rho t}\left(\cosh + \frac{\sigma}{q}\sinh\right) qt \cdot H_0, \\ E &= -\epsilon^{-\rho t}\frac{\sinh qt}{q}\frac{\nabla}{c} H_0, \end{aligned}\right\} \quad . \quad . \quad (34)$$

by (17). Let H_0 be zero on the right side and constant on the left side of the origin, and let us find H and E at a point on the right side. The operator $\epsilon^{vt\nabla}$ is inoperative, so that, by (30),

$$\left.\begin{aligned} H &= \tfrac{1}{2}\epsilon^{-\rho t}\epsilon^{-vt\nabla}(1 - sf_1 + s^2f_2 - s^3f_3 + \ldots)H_0, \\ E &= \tfrac{1}{2}\epsilon^{-\rho t}\epsilon^{-vt\nabla}(1 - sg_1 + s^2g_2 - s^3g_3 + \ldots)H_0 \times \mu v, \end{aligned}\right\} \quad . \quad (35)$$

the immediate integration of which gives

* These f's are the same as in my paper "On Electromagnetic Waves,' 8, Phil. Mag. February 1888; but s there is σ here.

$$\left.\begin{aligned} H &= \tfrac{1}{2}H_0\epsilon^{-\rho t}\left\{1+\sigma t f_1\left(1-\frac{z}{vt}\right)+\frac{(\sigma t)^2}{\underline{|2}}f_2\left(1-\frac{z}{vt}\right)^2+\ldots\right\}, \\ E &= \tfrac{1}{2}\mu v H_0\epsilon^{-\rho t}\left\{1+\sigma t g_1\left(1-\frac{z}{vt}\right)+\frac{(\sigma t)^2}{\underline{|2}}g_2\left(1-\frac{z}{vt}\right)^2+\ldots\right\}. \end{aligned}\right\} \quad (3$$

To obtain the E due to E_0 constant from $z=-\infty$ to 0, use the first of (36); change H to E, H_0 to E_0, and change the sign of σ, not forgetting in the f's. To obtain the corresponding H due to E_0, use the second of (36); change E to H, H_0 to E_0, and μ to c. So

$$\left.\begin{aligned} E &= \tfrac{1}{2}E_0\epsilon^{-\rho t}\left\{1-\sigma t f'_1\left(1-\frac{z}{vt}\right)+\frac{(\sigma t)^2}{\underline{|2}}f'_2\left(1-\frac{z}{vt}\right)^2-\ldots\right\}, \\ H &= \tfrac{1}{2}cvE_0\epsilon^{-\rho t}\left\{1+\sigma t g_1\left(1-\frac{z}{vt}\right)+\frac{(\sigma t)^2}{\underline{|2}}g_2\left(1-\frac{z}{vt}\right)^2+\ldots\right\}, \end{aligned}\right\} \quad (3$$

where the accent means that the sign of σ is changed in the f's.

From these, without going any further, we can obtain a general idea of the growth of the waves to the right and left of the origin, because the series are suitable for small values of σt. But, reserving a description till later, notice that E in (36) and H in (37) must be true on both sides of the origin; on expanding them in powers of z we consequently find that the coefficients of the odd powers of z vanish, by the first of (28), and what is left may be seen to be the expansion of

$$E=\tfrac{1}{2}\mu v H_0\epsilon^{-\rho t}J_0\left[\frac{\sigma}{v}(z^2-v^2t^2)^{\frac{1}{2}}\right], \quad . \quad . \quad (38)$$

the complete solution for E due to H_0. Similarly,

$$H=\tfrac{1}{2}cvE_0\epsilon^{-\rho t}J_0\left[\frac{\sigma}{v}(z^2-v^2t^2)^{\frac{1}{2}}\right] \quad . \quad . \quad . \quad (39)$$

is the complete solution for H due to E_0. In both cases the initial distribution was on the left side of the origin; but, if its sign be reversed, it may be put on the right side, without altering these solutions.

Similarly, by expanding the first of (36) and first of (37) in powers of z we get rid of the even powers of z, and produce the solutions given by me in a previous paper*, which, however, it is needless to write out here, owing to the complexity.

13. *Arbitrary Initial States.*—Knowing the solutions due to the above distributions, we find those due to initial E_0da at the origin, or H_0da, by differentiation to z; and for this we do not need the firsts of (36) and (37) but only the seconds.

* "Electromagnetic Waves," § 8 (Phil. Mag. Feb. 1888).

The results bring the Fourier integrals (21) to

$$\left.\begin{aligned} \mathrm{E}&=\epsilon^{-\rho t}\Big[\tfrac{1}{2}(\mathrm{E}_0+\mu v\mathrm{H}_0)_{z-vt}+\tfrac{1}{2}(\mathrm{E}_0-\mu v\mathrm{H}_0)_{z+vt} \\ &\qquad +\tfrac{1}{2}\int_{z-vt}^{z+vt}\Big\{\mathrm{E}_0\frac{-\sigma+p}{v}-\frac{\mathrm{H}_0}{cv}\nabla\Big\}\mathrm{J}_0(y)\,da\Big], \\ \mathrm{H}&=\epsilon^{-\rho t}\Big[\tfrac{1}{2}(\mathrm{H}_0+cv\mathrm{E}_0)_{z-vt}+\tfrac{1}{2}(\mathrm{H}_0-cv\mathrm{E}_0)_{z+vt} \\ &\qquad +\tfrac{1}{2}\int_{z-vt}^{z+vt}\Big\{\mathrm{H}_0\frac{\sigma+p}{v}-\frac{\mathrm{E}_0}{\mu v}\nabla\Big\}\mathrm{J}_0(y)\,da\Big]; \end{aligned}\right\} \quad . \ (40)$$

where

$$p=d/dt, \quad \nabla=d/dz, \quad y=\frac{\sigma}{v}\{(z-a)^2-v^2t^2\}^{\frac{1}{2}}.$$

Another interesting form is got by the changes of variables

$$\left.\begin{aligned} \mathrm{U}&=\tfrac{1}{2}\epsilon^{\rho t}(\mathrm{E}-\mu v\mathrm{H}), & u&=z-vt, \\ \mathrm{W}&=\tfrac{1}{2}\epsilon^{\rho t}(\mathrm{E}+\mu v\mathrm{H}), & w&=z+vt. \end{aligned}\right\} \quad . \quad . \quad (41)$$

These lead to

$$\left.\begin{aligned} \mathrm{U}_{z,t}&=\mathrm{U}_{w,0}+\int_u^w\Big(\mathrm{U}_0\frac{d}{dw}-\frac{\sigma}{2v}\mathrm{W}_0\Big)\mathrm{J}_0\Big\{\frac{\sigma}{v}(u-a)^{\frac{1}{2}}(w-a)^{\frac{1}{2}}\Big\}da, \\ \mathrm{W}_{z,t}&=\mathrm{W}_{u,0}-\int_u^w\Big(\mathrm{W}_0\frac{d}{du}+\frac{\sigma}{2v}\mathrm{U}_0\Big)\mathrm{J}_0\Big\{\frac{\sigma}{v}(u-a)^{\frac{1}{2}}(w-a)^{\frac{1}{2}}\Big\}da. \end{aligned}\right\} \quad (42)$$

The connexions and partial characteristic of U or W are

$$\frac{d\mathrm{W}}{dw}=-\frac{\sigma}{2v}\mathrm{U}, \quad \frac{d\mathrm{U}}{du}=+\frac{\sigma}{2v}\mathrm{W}, \quad \frac{d^2\mathrm{U}}{du\,dw}=-\Big(\frac{\sigma}{2v}\Big)^2\mathrm{U}; \ (43)$$

and this characteristic has a solution

$$\mathrm{U}=\Big(\frac{z+vt}{z-vt}\Big)^{\frac{m}{2}}\mathrm{J}_m\Big[\frac{\sigma}{v}(z^2-v^2t^2)^{\frac{1}{2}}\Big], \quad . \quad . \quad . \quad (44)$$

where m is any + integer, and in which the sign of the exponent may be reversed. We have utilized the case $m=0$ only.

14. *Evaluation of* Fourier *Integrals.*—The effectuation of the integration (direct) of the original Fourier integrals will be found to ultimately depend upon

$$\frac{2}{\pi}\int_0^\infty \cos mz\frac{\sinh qt}{q}\,dm=\frac{1}{v}\mathrm{J}_0\Big[\frac{\sigma}{v}(z^2-v^2t^2)^{\frac{1}{2}}\Big], \quad . \quad (45)$$

provided $vt>z$, where, as before,

$$q^2=\sigma^2-m^2v^2.$$

By equating coefficients of powers of z^2 in (45) we get

$$\frac{2}{\pi}\int \frac{\sinh qt}{q} m^{2r} dm = \frac{1 \cdot 3 \cdot 5 \cdot (2r-1)}{v^{2r+1}} \frac{\mathrm{J}_r(\sigma ti)}{(\sigma ti)^r}, \quad . \quad (46)$$

except with $r=0$; then

$$= \frac{1}{v} \mathrm{J}_0(\sigma ti).$$

To prove (45), expand the q function in powers of σ^2. Thus, symbolically written,

$$\frac{\sinh qt}{q} = \epsilon^{\frac{1}{2}\sigma^2 p^{-1} t}\left(\frac{\sin mvt}{mv}\right), \quad . \quad . \quad . \quad . \quad (47)$$

the operand being in the brackets, and p^{-1} meaning integration from 0 to t with respect to t. Thus, in full,

$$\frac{2}{\pi}\int \cos mz \frac{\sinh qt}{q} dm = \frac{2}{\pi}\int_0^\infty \cos mz \left[\frac{\sin mvt}{mv} + \frac{\sigma^2}{2}\int_0^t t \frac{\sin mvt}{mv} dt \right.$$
$$\left. + \frac{\sigma^2}{2}\frac{\sigma^2}{4}\int_0^t t\,dt \int_0^t t\,dt \frac{\sin mvt}{mv} + \dots \right] dm. \quad . \quad (48)$$

Now the value of the first term on the right is

$$\frac{1}{v}, \quad \frac{2}{v}, \quad \text{or } 0,$$

when z is $<$, $=$, or $> vt$.

Thus, in (48) if $z > vt$, since first term vanishes, so do all the rest, because their values are deduced from that of the first by integrations to t, which during the integrations is always $< z/v$. Therefore the value of the left number of (45) is zero when $z > vt$. In another form, disturbances cannot travel faster than at speed v.

But when $z < vt$ in (48), it is clear that whilst t' goes from 0 to t or from 0 to z/v, and then from z/v to t, the first integral is zero from 0 to z/v, so that the part z/v to t only counts. Therefore the second term is

$$\frac{2}{\pi}\frac{\sigma^2}{2}\int \cos mz \left[\int_{\frac{z}{v}}^t \frac{t \sin mvt}{mv} dt\right] dm = \frac{2}{\pi}\frac{\sigma^2}{2}\int_{\frac{z}{v}}^t t\,dt \int_0^\infty \cos mz \frac{\sin mvt}{mv} dm$$
$$= \frac{\sigma^2}{2}\frac{1}{v}\int_{\frac{z}{v}}^t t\,dt = \frac{1}{v}\frac{\sigma^2}{2^2}\left(t^2 - \frac{z^2}{v^2}\right).$$

The third is, similarly,

$$\frac{1}{v}\frac{\sigma^2}{2^2}\frac{\sigma^2}{4}\int_{\frac{z}{v}}^t t\left(t^2 - \frac{z^2}{v^2}\right) dt = \frac{1}{v}\frac{\sigma^4}{2^2 4^2}\left(t^2 - \frac{z^2}{v^2}\right)^2;$$

and so on, in a uniform manner, thus proving that the successive terms of (48) are the successive terms of the expansion of (45) (right number) in powers of σ^2; and therefore proving (45).

The following formulæ occur when the front of the wave is in question, where caution is needed in evaluations:—

$$\cosh \sigma t - \tfrac{1}{2} = \frac{2}{\pi}\int_0^\infty \frac{\sin mvt}{m} \cosh qt \, dm, \quad . \quad . \quad (49)$$

$$\frac{\sinh \sigma t}{\sigma} = \frac{2}{\pi}\int_0^\infty \frac{\sin mvt}{m} \frac{\sinh qt}{q} dm. \quad . \quad . \quad (50)$$

15. *Interpretation of Results.*—Having now given a condensation of the mathematical work, we may consider, in conclusion, the meaning and application of the formulæ. In doing so, we shall be greatly assisted by the elementary theory of a telegraphic circuit. It is not merely a mathematically analogous theory, but is, in all respects save one, essentially the same theory, physically, and the one exception is of a remarkable character. Let the circuit consist of a pair of equal parallel wires, or of a wire with a coaxial tube for the return, and let the medium between the wires be slightly conducting. Then, if the wires had no resistance, the problem of the transmission of waves would be the above problem of plane waves in a real dielectric, that is, with constants μ, c, and k, but without the magnetic conductivity; i. e. $g=0$ in the above.

The fact that the lines of magnetic and electric force are no longer straight is an unessential point. But it is, for convenience, best to take as variables, not the forces, but their line-integrals. Thus, if V be the line-integral of E across the dielectric between the wires, V takes the place of E. Thus kE, the density of the conduction-current, is replaced by KV, where K is the conductance of the dielectric per unit length of circuit, and $c\mathrm{E}/4\pi$, the displacement, becomes SV, where S is the permittance per unit length of circuit. The density of electric current $cp\mathrm{E}/4\pi$ is then replaced by SpV. Also SV is the charge per unit length of circuit.

Next, take the line-integral of $\mathrm{H}/4\pi$ round either conductor for magnetic variable. It is C, usually called the current in the wires. Then μH, the induction, becomes LC; where LC is the momentum per unit length of circuit, L being the inductance, such that $\mathrm{LS}v^2 = \mu c v^2 = 1$.

A more convenient transformation (to minimize the trouble with 4π's) is

E to V, H to C, μ to L, c to S, $4\pi k$ to K.

Now, lastly, the wires have resistance, and this is without any representation whatever in a real dielectric. But, as I have before shown, the effect of the resistance of the wires in attenuating and distorting waves is, to a first approximation (ignoring the effects of imperfect penetration of the magnetic field into the wires), representable in the same manner exactly as the corresponding effects due to g, the hypothetical magnetic conductivity of a dielectric*. Thus, in addition to the above,

$$4\pi g \text{ becomes } \mathrm{R},$$

R being the resistance of the circuit per unit length.

16. In the circuit, if infinitely long and perfectly insulated, the total charge is constant. This property is independent of the resistance of the wires. If there be leakage, the charge Q at time t is expressed in terms of the initial charge Q_0 by

$$\mathrm{Q}=\mathrm{Q}_0\epsilon^{-\mathrm{K}t/\mathrm{S}},$$

independent of the way the charge redistributes itself.

In the general medium, the corresponding property is persistence of displacement, no matter how it redistributes itself, provided k be zero, whatever g may be. And, if there be electric conductivity,

$$\int_{-\infty}^{\infty}\mathrm{D}\,dz=\left(\int_{-\infty}^{\infty}\mathrm{D}_0\,dz\right)\epsilon^{-4\pi kt/c},$$

where D_0 is the initial displacement, and D that at time t, functions of z.

In the circuit, if the wires have no resistance, the total momentum remains constant, however it may redistribute itself. This is an extension of Maxwell's well-known theory of a linear circuit of no resistance. The conductivity of the dielectric makes no difference in this property, though it causes a loss of energy. When the wires have resistance, then

$$\int_{-\infty}^{\infty}\mathrm{LC}\,dz=\left(\int_{-\infty}^{\infty}\mathrm{LC}_0\,dz\right)\epsilon^{-\mathrm{R}t/\mathrm{L}}$$

expresses the subsidence of total momentum; and this is independent of the manner of redistribution of the magnetic field and of the leakage.

In the general medium, when real, the corresponding property is persistence of the induction (or momentum); and when g is finite,

$$\int_{-\infty}^{\infty}\mu\mathrm{H}\,dz=\left(\int_{-\infty}^{\infty}\mu\mathrm{H}_0\,dz\right)\epsilon^{-4\pi gt/\mu}.$$

* "Electromagnetic Waves," § 6 (Phil. Mag. Feb. 1888).

In passing, I may remark that, in my interpretation of Maxwell's views, it is not his vector-potential **A**, the so-called electrokinetic momentum, that should have the physical idea of momentum associated with it, but the magnetic induction **B**. To illustrate, consider Maxwell's theory of a linear circuit of no resistance, the simplest case of persistence of momentum. We may express the fact by saying that the induction through the circuit remains constant, or that the line-integral of **A** along or in the circuit remains constant. These are perfectly equivalent. Now if we pass to an infinitely small closed circuit, the line-integral of **A** becomes **B** itself (per unit area). But if we consider an element of length only, we get lost at once.

Again, the magnetic energy being associated with **B**, (and **H**), so should be the momentum.

Suppose also we take the property that the line-integral of $-\dot{\mathbf{A}}$ is the E.M.F. in a circuit, and then consider $-\dot{\mathbf{A}}$ as the electric force of induction at a point. Its time-integral is **A**. But this is an electromotive impulse, not momentum.

Lastly, whilst **B** (or **H**) defines a physical property at a point, **A** does not, but depends upon the state of the whole field, to an infinite distance. In fact it sums up, in a certain way, the effect which would arise at a point from disturbances coming to it from all parts of the field. It is therefore, like the scalar electric potential, a mathematical concept merely, not indicative in any way of the actual state of the medium anywhere.

The time-integral of **H**, whose curl is proportional to the displacement, has equal claims to notice as a mathematical function which is of occasional use for facilitating calculations, but which should not, in my opinion, be elevated to the rank of a fundamental quantity, as was done by Maxwell with respect to **A**.

Independently of these considerations, the fact that **A** has often a scalar potential parasite, and also the other function, causes sometimes great mathematical complexity and indistinctness; and it is, for practical reasons, best to murder the whole lot, or at any rate merely employ them as subsidiary functions.

17. Returning to the telegraph-circuit, let the initial state be one of uniform V on the whole of the left side of the origin, $V=0$ on the right side, and $C=0$ everywhere. The diagram will serve to show roughly what happens in the three principal cases.

First of all we have ABCD to represent the curve of V_0, the origin being at C. When the disturbance has reached Z, that is when $t=\mathrm{CZ}/v$, the curve is A 1 1 1 1 Z f there be no leakage, when R and L are such that $\epsilon^{-\mathrm{R}t/2\mathrm{L}}=\frac{1}{2}$. At the

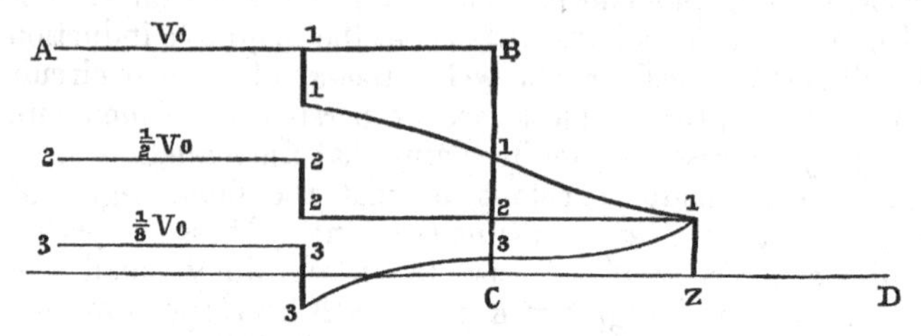

origin $\mathrm{V}=\frac{1}{2}\mathrm{V}_0$, at the front $\mathrm{V}=\frac{1}{4}\mathrm{V}_0$, and at the back $\mathrm{V}=\frac{3}{4}\mathrm{V}_0$.

Now introduce leakage to make $\mathrm{R/L}=\mathrm{K/S}$. Then 2 2 2 2 1 Z shows the curve of V, provided $\epsilon^{-\mathrm{K}t/\mathrm{S}}=\frac{1}{2}$. We have $\mathrm{V}=\frac{1}{2}\mathrm{V}_0$ on the left, and $\mathrm{V}=\frac{1}{4}\mathrm{V}_0$ in the rest.

Thirdly, let the leakage be in excess. Then, when V_0 has fallen, by leakage only, to $\frac{1}{8}\mathrm{V}_0$ on the left, the curve 3 3 3 3 1 Z shows V; it is $\frac{1}{16}\mathrm{V}_0$ at the origin, $-\frac{1}{8}\mathrm{V}_0$ at the back, and $\frac{1}{4}\mathrm{V}_0$ at the front.

Of course there has to be an adjustment of constants to make $\epsilon^{-\left(\frac{\mathrm{R}}{2\mathrm{L}}+\frac{\mathrm{K}}{2\mathrm{S}}\right)t}$ be the same $\frac{1}{2}$ in all cases, viz. the attenuation at the front.

18. Precisely the same applies when it is C_0 that is initially given instead of V_0, provided we change the sign of σ. That is, we have the curve 1 when the leakage is in excess, and the curve 3 when the leakage is smaller than that required to produce nondistortional transmission.

19. Now transferring attention to the general medium, if we make the substitution of magnetic conductivity for the resistance of the wires, the curve 1 would apply when it is E_0 that is the initial state and g in excess, and 3 when it is deficient; whilst if H_0 is the initial state, 1 applies when g is deficient, and 3 when in excess. But g is really zero, so we have the curve 1 for that of H and 3 for that of E.

This forcibly illustrates the fact that the diffusion of charge in a submarine cable and the diffusion of magnetic disturbances in a good conductor, though mathematically analogous, are physically quite different. They are both extreme cases of the same theory; but they arise by going to opposite extremities; with the peculiar result that, whereas the time-constant of retardation in a submarine cable is proportional to the resistance of the wire, that in the wire itself is proportional to its conductivity.

20. Going back to the diagram, if we shift the curves bodily

through unit distance to the left, and then take the difference between the new and the old curves, we shall obtain the curves showing how an initial distribution of V or C through unit-distance at the origin divides and spreads. In the case of curve 2, we have clean splitting without a trace of diffusion. In the other cases there is a diffused disturbance left behind between the terminal waves, positive in case 1, negative in case 3. But I have sufficiently described this matter in a former paper*.

October 18, 1888.

POSTSCRIPT.

On the Metaphysical Nature of the Propagation of the Potentials.

At the recent Bath Meeting of the British Association there was considerable discussion† in Section A on the question of the propagation of electric potential. I venture therefore to think that the following remarks upon this subject may be of interest.

According to the way of regarding the electromagnetic quantities I have consistently carried out since January 1885, the question of the propagation of, not merely the electric potential Ψ, but the vector potential **A**, does not present itself as one for discussion; and, when brought forward, proves to be one of a metaphysical nature.

We make acquaintance, experimentally, not with potentials, but with forces, and we formulate observed facts with the least amount of hypothesis, in terms of the electric force **E** and magnetic force **H**. In Maxwell's development of Faraday's views, **E** and **H** actually represent the state of the medium anywhere. (It comes to the same thing if we consider the fluxes, but less conveniently in general.) Granting this, it is perfectly obvious that in any case of propagation, since it is a physical state that is propagated, it is **E** and **H** that are propagated.

Now, in a limited class of cases, **E** is expressible as $-\nabla\Psi$. Considerations of mathematical simplicity alone then direct the mathematician's attention to Ψ and its investigation, rather than to that of **E** directly. But when this is possible the field is steady, and no question of propagation presents itself

* "Electromagnetic Waves," § 7 (Phil. Mag. Feb. 1888).

† See Prof. Lodge's "Sketch of the Electrical Papers read in Section A," the 'Electrician,' September 21 and 28, 1888.

(except in the very artificial form of balanced exchanges). When there is propagation, and **H** is involved, we have

$$\mathbf{E} = -\nabla\Psi - \dot{\mathbf{A}}.$$

Now this is, not an electromagnetic law specially, but strictly a truism, or mathematical identity. It becomes electromagnetic by the definition of **A**,

$$\text{curl } \mathbf{A} = \mu\mathbf{H},$$

leaving **A** indeterminate as regards a diverging part, which, however, we may merge in $-\nabla\Psi$. Supposing, then, **A** and Ψ to become fixed in this or some other way, the next question in connexion with propagation is, Can we, instead of the propagation of **E** and **H**, substitute that of Ψ and **A**, and obtain the same knowledge, irrespective of the artificiality of Ψ and **A**? The answer is perfectly plain—we cannot do so. We could only do it if Ψ, **A**, given everywhere, found **E** and **H**. But they cannot. **A** finds **H**, irrespective of Ψ, but both together will not find **E**. We require to know a third vector, $\dot{\mathbf{A}}$. Thus we have Ψ, **A**, and $\dot{\mathbf{A}}$ required, involving *seven* scalar specifications to find the *six* in **E** and **H**. Of these three quantities, the utility of **A** is simply to find **H**, so that we are brought to a highly complex way of representing the propagation of **E** in terms of Ψ and $\dot{\mathbf{A}}$, giving no information about **H**, which is, it seems to me, as complex and artificial as it is useless and indefinite.

Again, merely to emphasize the preceding, the variables chosen should be capable of representing the energy stored. Now the magnetic energy may be expressed in terms of **A**, though with entirely erroneous localization; but the electric energy cannot be expressed in terms of Ψ. Maxwell (chap. xi. vol. ii.) did it, but the application is strictly limited to electrostatics; in fact, Maxwell did not consider electric energy comprehensively. The full representation in terms of potentials requires Ψ and **Z**, the vector-potential of the magnetic current. [This is developed in my work "On Electromagnetic Induction and its Propagation," Electrician, 1885.] This inadequacy alone is sufficient to murder Ψ and **A**, considered as subjects of propagation.

Now take a concrete example, leaving the abstract mathematical reasoning. Let there be first no **E** or **H** anywhere. To produce any, impressed force is absolutely needed. Let it be impressed **e**, and of the simplest type, viz. an infinitely extended plane sheet of **e** of uniform intensity, acting normally to the plane. What happens? Nothing at all. Yet the potential on one side of the plane is made greater by the amount e (tensor of **e**) than on the other side. Say $\Psi = \frac{1}{2}e$ and $-\frac{1}{2}e$. Thus we have *instantaneous* propagation of Ψ to infinity. I prefer, however, to say that this is only a mathematical fiction, that nothing is propagated at all, that the electromagnetic mechanism is of such a nature that the applied forces are balanced *on the spot*, that is, in the sheet, by the reactions.

To emphasize this again, let the sheet be not infinite, but have a circular boundary. Let the medium be of uniform inductivity μ, and permittivity c. Then, irrespective of its conductivity, disturbances are propagated at speed $v = (\mu c)^{-\frac{1}{2}}$, and their source is the vortex-line of **e**, on the edge of the disk. At any time t less than a/v, where a is the radius of the disk, the disturbance is confined within a ring whose axis is the vortex-line. Everywhere else **E** $=0$ and **H** $=0$. On the surface of the ring, $\mathrm{E} = \mu v \mathrm{H}$, and **E** and **H** are perpendicular; there can be no normal component of either.

Now we can naturally explain the absence of any flux in the central portion of the disk by the applied forces being balanced by the reactions on the spot, until the wave arrives from the vortex-line. But how can we explain it in terms of Ψ, seeing that Ψ has now to change by the amount e at the disk, and yet be continuous everywhere else outside the ring? We cannot do it, so the propagation of Ψ fails altogether. Yet the actions involved must be the same whether the disk be small or infinitely great. We must therefore give up the idea altogether of the propagation of a Ψ to balance impressed force. In the ring itself, however, we may regard the propagation of Ψ (a different one), **A**, and $\dot{\mathbf{A}}$; or, more simply, of **E** and **H**.

If there be no conductivity, the steady electric field is assumed anywhere the moment the two waves from opposite ends of a diameter of the disk coexist; that is, as soon as the wave arrives from the more distant end (Phil. Mag. May 1888*). But this simplicity is quite exceptional, and seems to be confined to plane and spherical waves. In general there is a subsidence to the steady state after the initial phenomena.

If it be remarked that incompressibility (or something equivalent or resembling it) is needed in order that the medium may behave as described (*i. e.* no flux except at the vortex-line initially), and that if the medium be compressible we shall have other results (a pressural wave, for example, from the disk generally), the answer is that this is a wholly inde-

* "Electromagnetic Waves," § 25.

pendent matter, not involved in Maxwell's dielectric theory, though perhaps needing consideration in some other theory. But the moment we let the electric current have divergence (the absence of which makes the vortex-lines of **e** to be the sources of disturbances), we at once (in my experience) get lost in an almost impenetrable fog of potentials. Maxwell's theory unamended, on the other hand, works perfectly and without a trace of indefiniteness, provided we regard **E** and **H** as the variables, and discard his "equations of propagation" containing the two potentials.

October 22, 1888.

Printed in the United States
By Bookmasters